Satanic Mills or Silicon Islands?

SATANIC MILLS
or SILICON ISLANDS?

The Politics of High-Tech Production in the Philippines

STEVEN C. MCKAY

ILR PRESS AN IMPRINT OF

CORNELL UNIVERSITY PRESS, ITHACA AND LONDON

First published 2006 by Cornell University Press
First printing, Cornell Paperbacks, 2005
Printed in the United States of America

Library of Congress Cataloging-in-Publication Data

McKay, Steven C. (Steven Charles)
    Satanic mills or silicon islands? : the politics of high-tech production in the Philippines / Steven C. McKay.
        p. cm.
    Includes bibliographical references and index.
    ISBN-13: 978-0-8014-4236-0 (cloth : alk. paper)
    ISBN-10: 0-8014-4236-2 (cloth : alk. paper)
    ISBN-13: 978-0-8014-8894-8 (pbk : alk. paper)
    ISBN-10: 0-8014-8894-X (pbk : alk. paper)
    1. High technology industries—Philippines.   2. Electronic industry workers—Philippines.   3. Industrial relations—Philippines.   I. Title.
    HC460.H53M35   2005 2006
    331.7'62138109599—dc22
                                                                    2005025049

Cloth printing   10 9 8 7 6 5 4 3 2 1

Paperback printing   10 9 8 7 6 5 4 3 2 1

For my mother,
Remedios Bautista McKay

Contents

Acknowledgments

Few books are the products of political implosions and Plan B's. But in 1998, the volatile circumstances in Southeast Asia propelled my research in an unexpected yet fruitful arc, bringing me back to questions and to a country that first sparked a lifelong engagement with issues of justice and international development. Originally, I had planned a year of research in both Indonesia and the Philippines to compare the different labor politics and worker movements in the two Southeast Asian nations experiencing unprecedented growth and social change. But the financial crisis that broadsided the economies of the region also scuttled my initial comparative project.

Nevertheless, circumstances presented old and new questions about development, stability, and the role of foreign capital in an area that seemed to go from "boom" to "basket case" overnight. How does a more global, interconnected economy affect a locality's ability to transform itself? What roles do international investment and the nation state play in promoting or thwarting that transformation? And ultimately, what meanings and political possibilities does the "new global economy" open for those drawn into its web and who—by the sweat of their brows—bring it forth? Zeroing in on the Philippines and high-tech electronics refocused my attention on new state development strategies and the changing character of global production in which local workers find themselves deeply enmeshed.

The journey from asking such questions to the attempts to answer them in this book has been a long but enlightening one, aided by countless people and groups across two hemispheres. My first and greatest debt of gratitude goes to my mentor, advisor, cajoler, and friend, Gay Seidman, of the Department of Sociology at the University of Wisconsin, Madison, who provided intellectual guidance and critical feedback throughout the many stages of research and writing. Her scholarship, enthusiasm, and dedication to students are a model to which I humbly aspire. I also thank Russell Middleton, who has given his unwavering support and encouragement, and Jonathan Zeitlin and Jane Collins, for their engagement with the project and for asking the tough questions that only strengthened the work. Jamie Peck and Leann Tigges provided critical comments and intellectual inspiration. At UW Madison, I was fortunate enough to be affiliated with the Center for Southeast Asian Studies, where a core of leading Philippines scholars, namely Ian Coxhead, Michael Cullinane, Dan Doeppers, Paul Hutchcroft, Monita Manalo, and particularly, Alfred McCoy, offered tremendous training, guidance, and encouragement, and taught me much about what it means to be a dedicated and passionate researcher. I was also buoyed by the camaraderie and critical input of my fellow students and friends, including Kiko Benitez, Charlotte Kuo-Benitez, John Willauer, Carmel Capati, Dimitri Kessler, Jody Knauss, Vina Lanzona, Susan Mannon, Angel Adams Parham, Nancy Plankey Videla, Cynthia White, Jeff Rickert, Jeff Rothstein, Arthur Scarritt, Diane Soles, and Cliff Westfall.

My colleagues and the staff at the UW Milwaukee Department of Sociology—especially Carrie Costello, Kristin Espinosa, Laura Fingerson, Jennifer Jordan, Stacey Oliker, and Kent Redding—provided insightful comments on earlier drafts, collegial support, and the necessary laughs through the JFWC writing group. Philip Kelly, Michael Handel, Ligaya Lindio-McGovern, Alvin So, Steven Vallas, and Jean Wallace also provided helpful comments and suggestions on parts of the manuscript and at various panels and conference presentations.

Financial support for this project at various stages was provided by the Fulbright Institute for International Education, the U.S. Department of Education Foreign Language and Area Studies Fellowship, the MacArthur Foundation Consortium's Global Studies Program at the University of Wisconsin Madison, the Center for Southeast Asian Studies at UW Madison, and the University of Wisconsin, Milwaukee Graduate School.

In bringing this book to fruition, I was fortunate enough to work with Fran Benson at Cornell University Press, whose enthusiasm for the project was matched only by her patience and her editorial expertise. I would also like to thank the two anonymous reviewers who provided constructive critiques and keen insights that helped sharpen my arguments and improve the book's organization.

Of course, I am most indebted to the workers, managers, community members, and officials in the Philippines who gave generously and patiently of their limited time and taught me the profound complexities of work, development, and struggle from a myriad of perspectives. To all of you, *maraming salamat*. I would also like to thank the many people, institutions, and organizations that engaged my inquiries, helped open doors, and made the research possible. In particular I would like to thank the University of the

Philippines-Diliman School of Labor and Industrial Relations for providing research support and a vibrant intellectual community of scholars and staff, including Maragtas Amante, Cesario Azucena, Marie Aganon, Jose Gatchalian, Rene Ofreneo, Juan Amor Palafox, Melisa Serrano, Jorge Sibal, Virginia Teodosio, Hipolita Recalde, Roderico Amis, Lilian Cruz, and the rest of the SOLAIR faculty and staff. Thanks as well to Alex Calata at the Philippine-American Education Foundation, Gary Hawes at the Ford Foundation, and Richard Szal of the ILO-Manila. At the Department of Labor and Employment I acknowledge the help of Hector Morada of the Bureau of Labor and Employment Statistics, Teresa Soriano of the Institute of Labor Studies, and Benedicto Ernesto Bitonio and Alex Avela of the Bureau of Labor Relations. I gained additional insight and help from the field and central office staff of PEZA, Ernie Santiago at SEIPI, Tos Anonuevo at FES, Paul Quintos at EILER, Emilia Dapulang at NAFLU-NFL, Antonio Asper at FFW, Jurgette Hunculada at NCRFW, Boy Alonso, Father Lino De Castro, and Father Jose Dizon, Cecille Tucio, and Arnel Salvador of WAC. Special thanks to Venus Culili, Tess Dioquino, Liza Eleazar, and Elmer Piamonte for research and interview assistance.

My stay in the Philippines would not have been possible nor nearly as fulfilling if not for the warmth and generosity of my extended family. Florante and Purificacion Santos provided refuge from the Manila chaos, a real home in the cool hills of Tagaytay, and the joy of being part of their loving family. My cousins—Trixie, Edward, Ching, James, Jenny, and their families—also helped me in innumerable ways while making life in the Philippines a pleasure.

On less distant shores, I would like to thank my sisters and brother—Amy, Janie, and Ted—for the love, support, and playful ribbing that only older siblings can give. Thanks also to Bill and Linda Keys, for their generosity, steadfast sense of justice, and for making Madison such a wonderful place to live. And in acknowledging my parents, Anthony Portuguez McKay and Remedios Bautista McKay, who nurtured in me a curious and adventurous spirit, the arc has come full circle. Throughout my life, their unconditional love and unbridled encouragement has given me the confidence to believe that I can be anything, while leading me back to the Philippines to explore how I came to be. During a break from my research one day, I remember standing at the edge of the University of the Philippines campus, imagining a young Filipina graduate, some forty years before, facing a new, postwar world in a Philippines brimming with potential. This expectant, hopeful, and determined image repeated itself in the faces of the many young workers I interviewed. It is to this woman, my mother, that I dedicate this book.

Finally, my most heart-felt thanks and love to my own family, who have shared all my adventures and frustrations and have sacrificed so much to make this book—and me—possible. To Miles, for putting up with market sellers and an office door too often closed, but for embracing *tawilis,* typhoons, and silly dancing. To Charlie, who literally cut his teeth on mangos, Jollie Bee, and the Red Hot Chili Peppers. Most of all to Amy, my partner in crime, *abenteuer,* and love. For more years and on more shores than I can usually remember, you have been there, with your grace and strength and laugh, keeping the late-night coffee flowing and the spark incandescent.

Satanic Mills or Silicon Islands?

1 A New Politics of Production

AT 6:00 A.M., seventy-five bright blue buses, packed with dozing young women, career along scattered Philippine roadways and begin to converge. The rumbling buses throw dust clouds over rural roads and turn the heads of farmers already knee-deep in their fields. They funnel into the backstreets of a peri-urban boomtown clogged with belching jeepneys, squatters' shacks, and barbequed-banana stands. The acrid fumes of raw sewage and vulcanized tire repair choke the air. Out of the chaos and inevitable traffic jam, the buses turn into a restricted private road, past the reclining officers at the Provincial Industrial Police station, to finally reunite at the gates of the Philippine Technology Park.[1]

The guards wave the familiar buses into the empty streets of the export zone, an industrial park dotted with low-slung buildings sporting the logos of some of the world's best-known electronics companies. Arriving from strategic pickup points as far as two hours away, the buses shuttle some four thousand workers to the back door of Storage Limited, a manufacturer of bleeding-edge hard disk drives. A crowd of prospective applicants pooled around the guardhouse eyes the parade of buses enviously as it delivers the first shift.

Once inside, the women make their way to the long rows of lockers to wrestle into their hospital-blue "bunny suits." Talking and dressing in tandem, they tuck and straighten each other's full-body uniforms, covering each inch of skin until only their eyes are exposed. After swiping their ID badges into the electronic time clock, they file through the long air shower corridors, turning slowly with arms raised, before entering the virtually dust-free production area, a shop floor ten times more sterile than a hospital operating room.

1. All names of firms and zones in this study are pseudonyms.

The vast production floor, with its long, neat lines of automated assembly machines topped by red, yellow, and green warning lights, resembles an Orwellian casino. Yellow lines along the floor direct traffic. Through the production area's glass walls, passing managers glance at the color-coded target boards that hang above each section, announcing the productivity, quality, and defect rates tallied by the computer monitoring system. Signs suspended above the crowded testing area warn, "Absolute silence please." Video cameras hover in nearly every corner so those in the control room can watch, or not watch, what happens on the shop floor.

As they pour out of the air showers, workers take up the familiar positions they left just twelve hours before. Because of the hard disk drives' extreme sensitivity to static electricity, each worker wears a grounding cable attached to her body that she "plugs in" to her workstation. Workers may not leave their stations without permission, even to go to the bathroom. Thus, for most of their twelve-hour shift, six and often seven days a week, they stand literally tethered to the line.

Even outside the production area, workers must adhere to the strictly enforced "good housekeeping" policies that stress orderliness, discipline, and cleanliness in all areas of the plant. Roving inspectors, dubbed the "Quality Control Patrol," are armed with Polaroid cameras and photograph workers caught out of uniform or improperly dressed. They then display the pictures, with blacked out faces, on a large bulletin board outside the canteen as a warning to others.

Finally, an electronic bell chimes, the workers' muffled chatter dies away and the computerized assembly line lurches to life.

The scene described above conjures up contradictory images of the promise and menace of the global economy. Spurred by the dizzying changes in markets and technology, the trillion-dollar-a-year electronics sector has, since the 1990s, deluged developing countries with a wave of manufacturing investment. On the one hand, advanced electronics such as semiconductors and computer hard disk drives are widely coveted as leading industries in the twenty-first century. For receiving countries, such investments provide access to the latest technologies, bundled—it has been promised—with new ways to organize work, better jobs, and a survival strategy in the cutthroat world economy. On the other hand, the foregoing scenario seems to confirm our worst fears about globalization's impact on workers: that increased foreign investment only leads to higher levels of exploitation in low-cost production areas—the new "dark satanic mills"—where workers and nations are more vulnerable to the whims of unmoored capital.

In reality, neither of these polarized views captures the dynamic character of high-tech work. When I arrived in the Philippines to begin my fieldwork, I found the landscape of high-tech production vastly more complex. First, the multinational companies that have flooded into the Philippines are not simply assembling last year's boom boxes or robot pets. Rather, these firms churn out state-of-the-art products like the microprocessors in our wafer-thin laptops or the digital signal chips for our shrinking cell phones. This move to more sophisticated products has dramatically changed the organization of work, but not in uniform ways. Confounding the expectations of both critics and proponents of transnational production, I found—even within one industry in a single country—a wide array of curious organizational hybrids among firms, where production is flexible, innovative, and of high quality, yet can still remain quite "sweated."

Where production takes place has also changed. Although investments continue to pour almost exclusively into export processing zones, or EPZs, the zones themselves have been transformed. In the past, export processing zones, from Shenzhen to Central America, have been depicted by the media and academics as simply deregulated and deterritorialized spaces of cheap labor. But a new breed of zone has taken root in the Philippines, one owned by private developers who have strategically relocated and reorganized the zones, and downplayed the issue of wages. In fact, despite the relatively high cost of Filipino labor, the country now boasts one of the top five export processing zone programs in the world. In large part because of their revamped zones, the Philippines has become one of the few places in the world outside of China where leading electronics multinationals from the United States, Europe, and Asia have all set up manufacturing operations.

Finally, the workers themselves defied my expectations. At first glance, they seem to mirror the paradigmatic "nimble fingered" export processing zone worker found around the world: young women working exhausting schedules in tightly disciplined workplaces. But it seemed that reorganized production has strengthened the workers' hand: these workers are all permanent—rather than temporary—employees in heavily invested, high-profile multinationals that are acutely vulnerable to production disruptions. The Philippines also has a long history of militant labor unionism that led the fight for the world's first successful organizing of an entire export processing zone. Yet, in spite of often harsh working conditions and what looks to be an improved bargaining position, workers have shown little collective resistance. Indeed, many workers described their work to me as "clean," "easy," and even "fun," and seemed genuinely committed to their jobs and companies. Despite several concerted labor organizing campaigns, only about 1 percent of the labor force works under a union contract.

So how do we make sense of high-tech production as it emerges in developing countries like the Philippines? What explains the changes and wide variations in how work is organized? What can these changes tell us about the transformation of work in a globalizing economy? And finally, what consequences do these changes have for workers, the vast majority of whom are women? This book engages these key research questions by taking them up at a strategically crucial and empirically grounded flashpoint: a local site of global production and the local labor market in which it is embedded.

Specifically, through detailed case studies, I explore how four multinational electronics firms—American, Japanese, European, and Korean, respectively —located in four different Philippine export processing zones choose to organize work, control labor, and secure worker commitment by reaching beyond their factories and into workers' lives and communities. I also trace the rise of privatized export processing zones, highlighting the Philippine state's shifting but still significant role in labor control and the wider politics of global production. Finally, this book gives voice to the workers themselves, emphasizing the meanings they attach to high-tech work and their strategies to negotiate factory discipline on and beyond the "new" shop floor.

A common blind spot for both sides in globalization debates is a vision of globalization that is as frictionless as it is placeless; a vision in which developing countries appear as either simple sites of despotism or as substitutable templates for capital's unfolding production strategies. While globalization critics argue that low-cost production locations are becoming increasingly homogenized as they compete against one another to attract investment, proponents see high-tech production delivering the fruits of higher technology and better jobs anywhere. But at the core of both arguments is an overly generic characterization of production itself, whether exploitative or empowering, which makes the substitutability of locations possible. Overlooked is the elementary notion that, despite its transnational and high-tech nature, advanced production must still be constituted some*place,* by a specific group of workers in gritty locales like the one sketched above in the Philippines.

This book challenges the myths that high-tech production is generic and that globalization is simply a homogenizing force. First, I argue that technological change, new competitive demands, and contradictions within production are pushing high-tech firms to use a wide range of diverse organizational strategies, which I label flexible accumulation. Second, I argue that the restructuring of work has broadened the scope of labor control, extending it outside the factory and making the specificities and uniqueness of place *more,* not less, important. Third, I argue that firms, aided by host country governments, directly intervene in the local labor market in order to constitute and reproduce the social and gendered relations of flexible accumulation.

My arguments focus on two related processes integral to high-tech manufacturing in the Philippines. The first process is what I call the strategic localization of production. Here, I expand on a term used by economic geographer Andrew Mair (1997), referring to how firms develop site-specific labor control and work organization strategies depending on local patterns of labor regulation. The second process—a recasting of Michael Burawoy's (1979) notion of worker consent—is the active creation of worker commitment. Central to both of these processes and linking them together is the political construction of the local labor market. Each of these processes will be discussed in more detail below.

Most fundamentally, this book tells the story of how four different electronics manufacturing firms, facing contradictory demands in high-tech production and a crucial need for stability, craft a variety of new and more complex systems of production control. These variable systems include mastering new production technologies, securing worker commitment, and tapping into both formal and informal institutions of state regulation. The proper combinations of these elements are often realized through firms' localization strategies that exploit the uniqueness of place, the dynamics of space, and social cleavages in the labor force—based on gender, age, class, and residence—in order to leverage workers' labor market vulnerabilities and elicit worker commitment and consent. My objective is to look below macrolevel state interventions, beyond factory strategies, and more deeply into labor market manipulation to develop a broader, place-sensitive theory of high-tech production politics. This new model, which I call the "political apparatus of flexible accumulation," highlights the various roles that firms, state, and nonstate actors play in the reorganization of production and the reproduction of workers themselves.

Building a Theoretical Framework

In tackling the questions of how production is being restructured and how it affects workers and their lives, it is important to begin with a fundamental point: that work and markets are socially organized, and that the social relations in production depend heavily on the nature of managerial authority and the structures of organizational control. We start here because, while it has been the bedrock of much scholarship on work from Karl Marx to the labor process tradition to contemporary studies of organizations, the rise of restructured workplaces has prompted claims that "traditional" authority relations under mass production—often based on bureaucratically thick hierarchies—are being radically transformed in favor of devolved decision-making and increased worker autonomy and control over work.

Indeed, out of the ashes of what some see as a "failure" of centralized forms of mass production to respond to the new changes have emerged several models of restructured production that are viewed as not only more competitive, but also empowering for frontline workers, whose tacit knowledge and participation are seen as crucial. Framers of the first model, dubbed flexible specialization, argue that small, highly flexible, and networked firms using a strategy of permanent innovation, multiuse equipment, and a bevy of skilled production workers respond most ably to changing markets and technology (Piore and Sabel 1984). As an antidote to alienated labor under mass production, flexible specialization "is predicated on collaboration. . . . the production worker's intellectual participation in the work process is enhanced—and his or her role revitalized" (Piore and Sabel 1984, 278). A second and now more prevalent model is known as "lean production," which emerged from studies of the Japanese manufacturing industries and their emphasis on quality. The crucial elements here are new forms of efficiency—statistical process control, just-in-time production, low inventories, and "efficient" staffing—anchored, again, by the participation of multiskilled, securely tenured workers organized into teams (Kenney and Florida 1993; MacDuffie 1995; Womack et al. 1990). Finally, proponents of what are known as "high-performance work organizations" draw on the core technical and organizational elements of both previous models, but with a greater emphasis on the complementary human resource practices such as individualized incentive schemes and employment relations policies to inspire worker commitment and participation (Appelbaum et al. 2000; Belanger et al. 2002; MacDuffie 1995; Osterman 1994; Pfeffer 1994). Like their predecessors, these authors argue that firms adopting their model can become globally competitive precisely *because* they dismantle hierarchical relations in production and draw on workers' knowledge and decision-making autonomy to drive innovation and flexibility.

What the three strands of the optimistic flexibility literature have in common is a belief that mass production has failed as a viable mode of industrial organization. Ardent proponents claim that flexible or lean production is the new "one best way" to remain competitive under the new global capitalism. But while most manufacturers acknowledge that there have been changes in the demands of global competition, precious few are "taking the cure" of flexible production (Babson 1995; Osterman 2000). In the United States, despite dramatic results in several model cases such as the Saturn and NUMMI auto assembly plants, and the wide adoption of some practices across firms, the pace of innovation and organization change has been generally slow (Appelbaum and Batt 1994; Kalleberg 2003; Vallas 2003). According to Osterman's (2000) longitudinal survey of American firms, only about 15 percent of large firms have adopted fully reorganized workplaces and production processes.

So in the face of the ostensibly superior performance and competitiveness of restructured work organization, why have so few firms adopted these new models? The fundamental answer is that there are multiple paths to flexibility and ways to respond to changing economic conditions that do not *necessarily* call for the radical reorganization of production. This then leads to further questions: what has changed, and how are competitive firms responding, and their workers faring, under such changes? Here, it is important to begin with a reassessment of the core model from which so much of the flexible and lean production systems draw their inspiration; the "Japanese" system.

In a brilliant, historical account of the reception and adaptation of American management ideas in Japan, William Tsutsui (1998) argues that the success of contemporary Japanese industry is due in large part to the ways in which F. W. Taylor's scientific management ideas have worked both as an ideological framework and as a shop floor methodology. Tsutsui traces how the ideas of progress through rationality and science were extended from the shop floor to the industry and national political level to form the core of Japan's postwar productivist vision for economic growth. This productivist ideology served to enhance, and was itself undergirded by, the application and adaptation of engineer-led Taylorist practices, such as the appropriation of workers' craft knowledge and the application of statistical process control. Lean production at Toyota, he shows, relied particularly heavily on the core Taylorist practices of time-and-motion studies and layout design developed during the war and postwar period. Tsutsui (1998, 242–43) concludes:

> The pervasive influence of Taylorism and the frustrated career of Fordism suggest that a label like "non-Fordist"—or, better yet, "revised Taylorite"—best captures the nature of contemporary Japanese production arrangements. . . . In Japan as in the West, managerial commitment to the humanization of work has remained shallow and, for the most part, rhetorical rather than practical: even Japanese innovations like quality control circles, so widely praised as models of democratic and compassionate management, were born of the perennial managerial drive to Taylorize the shop floor.

When we demystify "Japanese-style management" and expose its essentially Taylorist roots, it becomes obvious that combining flexibility with hierarchical employee relations remains a viable and competitive strategy in advanced manufacturing. For example, in large apparel firms in both the United States and abroad, managers deploy expensive new flexible technologies to respond to the vicissitudes of a changing market, yet maintain the same, fundamentally labor-intensive production processes and leave unchanged hierarchical relations on the shop floor (Collins 2003; Taplin 1995). Similarly in the North American auto industry, lean production has not proven to be a blueprint, but rather more like a menu of technical and social practices (Babson 1999). Firms may choose from the menu depending on the specific technical requirements of the product, but more important, depending on "the particular history of the plant- or firm-level collective bargaining, past practices, demographics, and other factors [that] constrain or facilitate implementation" (Babson 1999, 26).

This critical approach is particularly relevant for lean production as it is applied in developing countries. Shaiken and others have shown convincingly that competitive, high-tech production can be successfully sited in a low-wage, low-skill context and in ways that do not fundamentally disturb traditional power relations on the shop floor (Shaiken and Browne 1991; Deyo 1997). Luthje (2004) describes state-of-the-art semiconductor plants in China that are flexible and produce high quality products, but in which work remains resolutely standardized and workers directly controlled by supervisors. In fact, it was found that managerial flexibility, more than low wages, proved more important to production efficiency, particularly in highly automated plants. This flexibility was gained primarily because management had a freer hand, unencumbered by union opposition (Shaiken 1995; see also Herzenberg 1996; Carrillo 1995).

Thus it becomes increasingly apparent that recent market changes are intensifying competition and pushing firms to adopt strategies, generally designed along neo-Taylorist lines, to produce more (and better) with less. However, this important but still quite general insight does not bring us closer to understanding why, in the Philippine case, there remains so much organizational variation between the high-tech firms in a single industry. Why do workers in these firms continue to be committed to such neo-Taylorist jobs? And what do the recent changes in production have to do with changes in how the export processing zones are organized? What is needed is a stronger theoretical framework for understanding both continuity and change under globalization, and the variation in the meanings and organization of work in advanced manufacturing.

Labor Control and the Politics of Production

In this book I propose we revisit the issues of labor control and the social relations in production that were at the core of labor process theory, but in light of recent trends, recast and transcend the analytical framework of one of its most advanced theorists, Michael Burawoy. Specifically, I develop a more comprehensive model of labor control, focusing on three key areas where the "new" politics of high-tech production play out, namely: at the point of production, where workers and mangers struggle over contradictory logics in the labor process; in the local labor market, where firms engage in localization strategies to enhance their bargaining position vis-à-vis labor; and finally, through national and local state institutions that regulate and help reproduce the social relations in and of production. Through this integrated approach, I hope to better explain the wide range of organizational strategies I found among the four case studies, the impetus for the reorganization of the export processing zones, and the reasons for the low levels of collective resistance among the high-tech manufacturing work force.

Burawoy (1985) breathed both controversy and new life into labor control debates by theorizing a dynamic link between shop floor politics and broader state politics. Based on his distinction between the *labor process*—or the technical organization of production—and the *political apparatuses of production*—or the institutions that regulate and shape workplace struggles—he posits two basic types of capitalist factory regimes: the despotic, in which work is organized around shop floor coercion; and the hegemonic, in which workers "must be persuaded to cooperate" or consent to their own exploitation (Burawoy 1985, 126). Although Burawoy identifies four contributing factors to regime difference—the labor process, market competition, labor power reproduction, and state intervention—it is state intervention that critically distinguishes despotic from hegemonic regimes. He argues that when states provide welfare benefits and enforce labor legislation, workers gain enough bargaining leverage to extract concessions from management, leading to a consent-based hegemonic regime. But he also warns that hegemonic regimes in advanced countries are undermined by capital mobility and competition from developing countries, where low-wage, despotic work regimes still reign.

While providing a strong analytical framework to approach restructuring, Burawoy's unmodified theory is at pains to explain the complexity of work organization and the blurring of coercion and consent in contemporary electronics manufacturing, particularly in developing countries. Burawoy, like many globalization critics, argues that production in developing country export processing zones simply exploits cheap labor, "requiring brutal coercion at the point of production" (Burawoy 1985, 265). Although such despotism still exists—as we will see in one of the case studies—production in the larger, now dominant sectors of advanced electronics manufacturing is not so neatly characterized.

Recent technological and market changes have meant increased capital intensity and a more complex competitiveness. Firms producing high-tech commodities such as hard disk drives, microprocessors, and integrated semiconductor chips now compete simultaneously on the bases of cost, quality, product differentiation, *and* speed-to-market (Ernst 2003). They must also orchestrate intricate production networks that are both globally dispersed and regionally agglomerated (McKendrick et al. 2000). For the labor process, these market imperatives create a number of contradictions. Manufacturers must juggle the standardizing pressures of high-quality, high-volume production and economies of scale with the flexibility pressures of extremely short product cycles, rapid technical innovation, and responsiveness to customers. These contradictory demands often mean that firms experiment with a variety of production organization and labor control strategies to achieve what might best be called competitive flexible production (Vallas 2003; Babson 1999). And as many studies have shown, work regimes based solely on coercion and simple control often lead to instability, inflexibility, and poor quality—precisely those areas in which advanced producers must compete (Edwards 1979; Burawoy 1985).

Given the need for production quality and stability, management often prefers worker consent over coercion, even in the absence of state intervention to enforce labor laws or provide welfare provisions. This is increasingly true for final assembly processes, since on-time delivery and relations with customers are crucial and a consent-based system is less disruptive. As will be shown in the case studies and discussed in detail below, management tries to organize worker commitment in order to reduce costly turnover and head off any stoppages or slowdowns in production. Firms must also contend with existing and potential collective worker resistance, employing different strategies depending on, as Babson (1999) noted above, the history of collective bargaining, past practices, worker demographics, and other factors that might restrain the free hand of management.

These new and more active strategies to stabilize production exploit a broader basis for constructing worker consent than Burawoy and others theorize. In his analysis, Burawoy remains firmly focused on the shop floor. Other theorists, such as Ching Kwan Lee (1998) in her comparative study of workers in Hong Kong and Guangdong, China, extend his insights, arguing that worker consent is crucial but that nonclass subjectivities formed outside the workplace—based on gender, culture, and conditions in the labor market—are equally important factors shaping worker interests. However, Lee, like Burawoy, fails to recognize alternative types of state intervention beyond the bureaucratic model that nevertheless have a direct role in reproducing labor power and constructing worker consent. Burawoy implies that the state acts primarily on the macroeconomic level: providing general welfare benefits and regulating industrial relations.[2] Lee cites only the state's conspicuous absence in intervening at the firm level. Yet increasingly, because of the complexities of global production and competition, states must often go beyond traditional bureaucratic regulation at the national level in order to draw and retain investment (O'Riain 2004).

Finally, in Lee's extended model of consent, it is workers' labor market dependence that "determines management strategies of incorporating labor" (Lee 1998, 12). However, her account treats conditions in the labor market as entirely exogenous. As noted above, she does not acknowledge the active role of the state, a central player in the political construction of the labor market. She also underestimates the power of employers, who do not simply respond to labor market conditions but can actively shape labor markets in order to increase worker dependence, enhance commitment, and diffuse resistance.

2. This position is consistent with neoinstitutionalists who recognize that states under globalization still matter, but mainly through national-level policies (Hollingsworth and Boyer 1997).

A New Framework for a New Politics

Changes in advanced manufacturing are leading to a greater need for production stability and worker consent. But both firms and the state have developed new strategies and intervene in a variety of arenas, constructing more complex political apparatuses of production than Burawoy considers. In particular, the active roles of both employers and the state in the manipulation of the labor market require a rethinking and expanded understanding of labor control and worker consent that goes beyond the shop floor.

Beginning in production, I draw on the theories of "flexible accumulation" as a better way to characterize and understand recent changes in work organization in the face of technological change, intensified global competition, and rising uncertainty since the 1970s (Harvey 1989; Rubin 1995; Gottfried 2000). At the macro level, flexible accumulation represents a regime shift in the dominant mode of economic growth, distribution, and regulation, from mass production, mass consumption, and a liberal welfare state under "Fordism" toward more customized production, fragmented markets, and increasingly neoliberal governance under "flexible accumulation" (Piore and Sabel 1984; Harvey 2001).[3] At the level of production, flexible accumulation represents a shift in the labor process and labor relations. As Vallas (1999, 91) explains in his critical review of workplace restructuring, "As firms added new and more flexible ways to accumulating capital— or put differently, removed inherited barriers to profitability—they have at the same time refashioned the structures of work, labor markets, and the employment relation itself." In direct contrast to theories of lean production or high-performance work organizations that see such changes in technology and organization as having positive or at least neutral impact on production workers, flexible accumulation theory views these refashioned structures as far less benign.

First, Vallas (1999; 2003) and others note that restructuring often *intensifies*, rather than diminishes the separation between "professional" and "production" workers. In the case studies to follow, this is most apparent in the growing divide between production operators and technical or engineering staff. Crucially, this process of technological change on the shop floor is also highly gendered, intensifying an already gender-segmented job structure and limiting the potential for mobility and upskilling for women production workers. Second, flexible accumulation has led to the rise of decentralized but hierarchical production networks and new power relations, such as the rise in customer power and the dominance of the parent

3. Gottfried (2000) makes the same distinction between modes of economic growth, but labels the "new" regime following Fordism as "Neo-Fordism" to highlight the continuity of Fordist, or standardized, hierarchical ways to organize work, rather than a fundamental break.

firm through research and design. In this book, the firms are all branch plants of much larger electronics multinationals involved primarily in assembly and test manufacturing. At this end in the production chain, their principal concern is not generating innovation through genuine worker participation, but primarily positive customer interaction and production stability through workplace control. These issues of the limits of workplace participation for worker empowerment and the widening (and gendered) gap between production and technical workers will be discussed more fully in chapter 3. Finally, flexible accumulation recognizes that firms not only consider, but can also affect the social and cultural environments in which they exist. While Vallas (1999) focuses on how firms influence tastes and market demand rather than treat them as exogenous, I focus in this research on how firms not only actively engage with their localities but can shape inequalities within them as well.

To better understand how work is organized in the four case studies, I develop a more integrated notion of labor control drawing from Burawoy's own emphasis on the *reproduction* of the social relations in and of production. Here I draw on the work of critical economic geographers for their insights into how differences or segmentation both within and between labor markets and locales interact with investment decisions and the organization of production itself. In terms of labor control, I take a more expansive, place-sensitive view following Peck's (1996, 179) broader definition: "labor control refers to the reproduction of the social relations of both the labor process and the labor market. This conception of labor control embraces the interrelated processes of (1) securing an appropriate labor supply, (2) maintaining control within the labor process and (3) reproducing this set of social relations."

My approach, then, is a reconstruction of Burawoy's political apparatuses of production modified for new forms of flexible accumulation, giving equal weight to the three interrelated processes of labor control. Specifically, I look below the macrolevel interventions of the national welfare state and the interests of "capital-in-general," evoking Jonas's (1996) concept of a local labor control regime. As Jonas (1996, 335) explains:

> Whereas capital-in-general is interested in the free and unlimited exchange of labor power, particular capitals are sensitive to the local contexts in which that exchange takes place. As such, they develop labor control strategies which limit the "freedom" of labor and regulate the conditions under which it enters the labor process.

Using the lens of local labor control, I examine two key processes of reproducing labor power and the work regimes taking place in the Philippines: the strategic localization of production and the active construction of worker commitment. I trace these two key processes as they shift across different local institutional domains, namely, in the factories, in the labor market, and through national and local governments.

Strategic Localization of Global Production

One of the central goals of this book is to demonstrate not only *that* localities continue to matter under globalization, but *how*. To this end, I develop an expanded notion of strategic localization to better understand the role of place in securing control over production and the consent of workers. Strategic localization can also help illuminate how both firms and local governments are intimately involved in constructing a multilayered, place-based political apparatus to regulate high-tech production. Despite the outcry from globalization critics, the dispersal of increasingly sophisticated production across space has not created a kind of placeless globalism in which all capital is necessarily footloose and production locations equally substitutable (Frobel et al. 1978; Burawoy 1985; Greider 1997). Rather, the new, more complex competitiveness has lead firms to seek distinct local "fixities" such as pockets of uniquely skilled labor and particular regulatory regimes to maximize their cost, quality, *and* speed-to-market advantages (O'Riain 2004; Brenner 2004). The specific nature of the firm, its production process, and its labor needs will influence what types of workers, and thus what type of locality, it seeks (Massey 1995). As we will see, the four multinational electronics firms in this book all came to the Philippines in the mid 1990s for a complex set of reasons that include the availability of highly educated and English-speaking production and technical workers, proximity to corporate headquarters and/or key markets, existing clusters of other high-tech producers, and a variety of government incentives aimed specifically at high-tech manufacturers.

National and local states, then, play crucial roles in strategic localization by attempting to lure such global investment. State actors matter not only because they preside over basic regulatory policies, such as setting wages, enforcing employment contracts, providing social welfare benefits, and crafting wider economic policies (Berger and Dore 1996; Hollingsworth and Boyer 1997). Increasingly, state actors at multiple levels respond directly to particular industrial investors by pursuing more targeted strategies such as upgrading complementary infrastructure and selectively regulating both industry and labor to provide the appropriate conditions for more advanced production (O'Riain 2004). To meet the changing needs of the electronics industry, the Philippine state has quite actively transformed its export processing zone program from an emphasis on deregulated public zones to attract simple manufacturing assemblers in the 1970s to a model of reregulated, privatized high-tech enclaves that appeal to the leading multinational manufacturers of the twenty-first century. The transformation of the zone program and its accommodation to the needs of the global electronics industry will be more fully explored in chapter 4.

Under strategic localization, multinational firms also actively adjust their policies and work regimes to fit local conditions. For example, Japanese consumer electronics producers that have set up production in California use very few "Japanese" production practices such as quality circles or JIT (Milkman 1991). Instead, these firms chose to adapt most of their policies to local conditions, norms, and practices. As Andrew Mair (1997, 80), studying the Honda Motor Corporation in the United States and Europe, notes: "The objective of local human resource managers was precisely to design structures able to build a *coherent bridge* linking Honda's fixed labor process requirements, such as very low absenteeism, good discipline, assignment flexibility, and willingness to suggest improvements, with the local cultural and social environment" (emphasis in original).

But what Mair refers to as "the local cultural and social environment" is in fact primarily patterns of local labor regulation and conditions and wages in local labor markets. And, as we will see in the four case studies, employers not only try to adapt to local conditions, but in fact also attempt to directly and indirectly influence conditions in the labor market to enhance their competitive advantage and control over labor and the work process. At the national level, for example, firms may seek exemptions from national labor laws in order to align location conditions to their work regimes—rather than the other way around. The weak bargaining position of developing country governments and their desperate need to attract investments often mean they must grant such concessions and refrain from imposing too many restrictions. But not all local conditions can be negotiated at the national level, and all firms must engage with their locality, particularly in terms of building and reproducing their workforce.

Manipulating Labor Markets

The clearest demonstration of how location continues to matter even after the initial investment is in the workings and regulation of local labor markets. Rather than accept an economist's human capital view of the labor market as a container for universal, smooth, and power-neutral market exchanges, I draw on more sociological theories of the labor market that recognize it as a power-laden and contingent process of negotiation between firms, workers, and their networks in matching different kinds of people with different types of jobs (Granovetter and Tilly 1988; Tilly and Tilly 1994). Employers are interested primarily in two main aspects of matching jobs and workers: productivity and organizational maintenance (Tilly and Tilly 1994). Organizational maintenance, in the context of advanced electronics production, is focused on the stability of production and the construction of worker commitment. But as we will see through the case studies, there are multiple ways to maximize both.

To understand how this can happen, I turn to labor market segmentation theory, which recognizes that the differences among people and among jobs are stratified—that is, the rules governing the matching process differ in each segment and that mobility between segments is often difficult (Peck 1996). Thus workers and job seekers are characterized and grouped in a wide variety of ways—by age, gender, race, civil status—giving them different levels of labor market bargaining power even if they share ostensibly equal "human capital." While employers do not necessarily create unequal social divisions, this does not prevent them from exploiting, and thus reproducing them. As Rubery and Wilkinson (1994, 31–32, emphasis added) note:

> Segmented labor markets with comparable labor available at different terms and conditions provide the opportunity to employers to tailor their labor market strategies to their needs without necessarily sacrificing the benefits of an established and committed workforce.... Thus employers have the best of several worlds: the domestic circumstances of married women, for example, provides the basis for a flexible, committed but cheap labor force: *primary workers at secondary prices.*

In the four cases I present in this book, we will see how different firms, with different production strategies, choose to manipulate the labor market positions of diverse groups through their gendered recruitment practices, union avoidance policies, and active disorganization and dispersal of worker housing.

It must also be emphasized that other actors and conditions affect this negotiation between workers and employers. As Miriam Wells (1996) points out, labor markets are not just socially but also politically constructed. In her study of the California strawberry industry, she highlights how active state policies, such as border control, immigration, and (lack of) labor protection shape labor markets and workers' labor market leverage even before employment. In particular, she demonstrates that the surplus labor market conditions can have the appearance of being unstructured and competitive. But, in fact, political forces and state policies often affect aggregate labor supply and constrain the options of labor market participants.

Similar forces affecting labor market power seem to be at work in the Philippines. Given the vital importance of political stability to multinational firms and the Philippine state's own dependence on foreign investment, the state plays a critical role in labor control and the politicization of the production context. As we shall see in chapter 4, the state actively reduces workers' bargaining power, even before they step onto the shop floor, through the reorganization and regulation of export processing zones (EPZs), enforcement of "industrial peace," and in its selective *non*enforcement of the Philippine Labor Code, which guarantees the right to organize and bargain collectively.

Securing Commitment in an Insecure World

The second key process that helps stabilize high-tech production and secure labor control is the active construction of worker commitment. My focus on securing commitment is akin to Burawoy's notion of "manufacturing consent," or the need to persuade workers to cooperate. However, as many have pointed out, Burawoy's understanding of consent and worker interests remains undertheorized because it fails to unravel, from a worker's point of view, all the myriad elements (gender, age, culture, and other experiences) that contribute to workers' interests, identities, and subjectivities—those that go beyond issues of class and beyond the shop floor (Lee 1998; Freeman 2000; Salzinger 2003). At the same time, traditional studies of commitment from the psychology, organizations, and human resource literatures remain woefully underpoliticized; these studies often neglect issues of power and control, the effectiveness of negative sanctions, and the influence of wider social inequalities. While proponents of "high performance" or "high commitment" work organizations are correct that worker commitment is crucial to advanced manufacturing, too often they assume a direct causal relationship between positive incentives, worker commitment, participation, discretionary effort, and improved firm performance (Appelbaum et al. 2000; MacDuffie 1995; Lincoln and Kalleberg 1990). As I will show, worker commitment is deeply intertwined with management strategies to maintain power and control: the four firms in this book elicit worker commitment primarily to ensure workplace stability, dampen worker disruption and turnover, and increase overall control over production. And in securing this commitment, slack and segmented labor markets again play a vital role, as firms leverage workers' labor market vulnerabilities and dependence on their jobs to induce greater willingness to accept conditions and terms at work.

At least since Hirschman's (1970) theory of exit, voice, and loyalty, research on organizational commitment has focused on a firm's internal and positive rewards system (see Meyer and Allen 1997 for a review). But while firms in this book do use positive internal incentives, such as good pay and benefits, they also use many negative and external strategies, such as strict shop floor discipline and selective recruiting of women workers from impoverished rural areas. To understand the use of both positive and negative incentives, we must break down and analyze organizational commitment as it relates to a firm's product and business strategy. Commitment is most commonly defined as the

> relative strength of an individual's identification with and involvement in a particular organization. Conceptually, it can be characterized by at least three factors: (a) a strong belief in and acceptance of the organization's goals and values; (b) a willingness to exert considerable effort on behalf of the organization; and (c) a strong desire to maintain membership in the organization. (Mowday, Steers, and Porter 1982, 27; see also Mueller and Wallace 1992)

The high commitment literature puts great emphasis on the first part of this definition, the acceptance of organizational goals and identification with the firm, also known as affective commitment or, simply, loyalty (Appelbaum et al. 2000; Lincoln and Kalleberg 1990; Mueller and Wallace 1992). Loyalty is important because it builds trust and drives workers to participate in problem solving (Lincoln and Kalleberg 1990; 1996). However, in the four firms in this book, loyalty and participatory problem solving were the *least* important aspects of firms' competitive and commitment strategies.

Rather, the second and third elements of commitment, which I call effort and attachment, prove both far more consequential and less reliant on internal, positive incentives. Effort—the willingness to work hard—is a central focus of firm strategies to increase productivity (Lincoln and Kalleberg 1990). But as has been emphasized in the long tradition of workplace literature, firms often ensure worker effort through means of direct, technical, and bureaucratic control (Braverman 1974; Edwards 1979; Herzenberg, Alic, and Wial 1998). A range of recent studies of high commitment workplaces has also documented the use of both positive *and negative* incentives to improve effort: from performance-based pay, job security, and employee participation to coercion, surveillance, and job enlargement (Kinnie, Hutchinson, and Purcell 2000; Kraft 1999; Lincoln and Kalleberg 1996; Smith 2001).

The third element—attachment—is often referred to as intent to stay and reflects firms' concerns with employee turnover (Mueller and Boyer 1994). Firms often use positive incentives to increase worker attachment; workers may be appreciative of incentives and/or view accumulated benefits as "side bets" that they do not want to lose by quitting (Becker 1960; Halaby and Weakliem 1989). However, previous research—particularly from those who focus on labor market segmentation—shows that attachment is also heavily influenced by external factors, such as lack of alternative work opportunities or family responsibilities (Mueller and Boyer 1994; Edwards 1979). Penley and Gould (1988) drawing on Etzioni (1975), develop the useful idea of alienative commitment. Seemingly contradictory, alienative commitment describes a condition in which workers feel a lack of control over work, yet because of a perceived absence of labor market alternatives, nevertheless remain attached to the firm (Penley and Gould 1988).

Thus contrary to much of the recent high commitment literature, not only is commitment affected by both positive and negative incentives, but all three components of commitment are also influenced by factors internal and external to the firm. As will be shown in the case studies, workers' subjectivities and commitment attitudes toward their jobs are shaped not only at the point of production, but are fundamentally influenced by other factors, such as gender, age, education, work experience, and labor market conditions. Introducing the notion of alienative commitment and relating commitment to the broader social context in which it is formed helps account for the different types of worker commitment found and the contradictory existence of high worker commitment despite tightly controlled working conditions.

But this not to say that workers gain no benefits from their jobs or simply work out of desperation. Commitment, while shaped by employers, is necessarily negotiated and ultimately must be given by workers themselves. Therefore it is important to recognize workers' agency and take into account workers' subjectivities, interests, and points of reference for decision-making. Particularly important in this regard are the issues of gender, power, and agency that are inextricably bound with the processes of restructuring and globalization. As Freeman (2000, 36) notes, "Global assembly lines are as much about the production of people and identities as they are about restructuring labor and capital." As will be discussed in chapter 5, for many women, work in the electronics industry, and especially with a large multinational firm, offers status, agency, and greater control over their lives. However, it must also be noted that this is to some extent the product of firms' selective and gendered recruiting in rural areas. Like recent migrants around the world, workers' assessment of how "good" a job is and their willingness to commit to it are often still based on conditions in their home communities—where opportunities are often limited and unemployment quite high. Thus while workers do experience real autonomy, this type of asymmetric agency does not necessarily challenge management authority at work nor disturb the power inequalities between workers and managers.

Industry and Case Selection

This book focuses on the high-tech electronics manufacturing sector in the Philippines for a number of reasons. Beginning in the late 1970s, many early debates about industrial organization focused on the electronics sector, especially the emergence of export processing zones, a "new international division of labor," and the impact of these phenomena on developing countries and particularly their women workers (Frobel et al. 1978; Lim 1990; Lin 1987; Fernandez-Kelly 1983; Elson and Pearson 1981; Aldana 1989; Ong 1987; Safa 1981; Sklair 1989). Yet by the 1990s, others began pointing to the same industry for its potential to transform the very nature of work. Some of the early and most ardent proponents of the lean production model cite high-technology electronics as the paradigmatic industry to assess the new forms of competitive work reorganization (Kenney and Florida 1993). Indeed, it is argued that electronics—particularly semiconductors and computer hard disk drives—is a key sector at the forefront of economic globalization because it has long faced the intense technological changes and competitive pressures that are viewed as harbingers of the twenty-first-century economy (Linden, Brown, and Appleyard 2001; Ernst 2003; McKendrick, Doner, and Haggard 2000). In some sense, I hope to revisit some of the early critical writings on work and women workers in the electronics sector in order to better appraise the impact of more recent changes in the organization of global production (Acker 2004; Chun 2001; Standing 1999; Freeman 2000; Salzinger 2003).

I focus on the Philippines in part because of its relatively long history and experience with export processing zones and investment by leading multinational electronics firms, both dating back to 1974. By the mid 1990s, the Philippines had emerged as a premier site for technology-intensive assembly and test manufacturing of semiconductors and computer hard disk drives (Tecson 1999). Some of the biggest industry names from the United States, Europe, and Asia are here, including Intel, Texas Instruments, Read-Rite, Seagate, Philips, Analog Devices, Fujitsu, Toshiba, TDK, Hitachi, and Acer. In fact, more than 70 percent of the more than eight hundred electronics firms are foreign-owned (SEIPI 2003). From 1994 to 2001 multinationals poured in nearly $9.5 billion, almost all of it for production facilities located within the forty-seven active Philippine economic zones. By 2004, the electronics exports exceeded $26 billion, or almost 70 percent of all Philippine exports, and the industry employed more than 346,000 workers (NSO 2005; DTI/BETP 2003).

My analysis centers on four case studies of firms and their workers that reveal four distinct work regimes, or ways to organize work, localize production, and control labor.[4] These case studies, compared and presented in rich detail in the chapters to follow, are summarized in table 1.1 below. As the table demonstrates, the work regimes vary considerably in terms of the

---

4. A work regime is much like a work system: consisting of production and human resource practices, work organization, and employment relations (Herzenberg, Alic and Wial 1998; Katz and Darbishire 2000). But a work regime extends the notion of a work system as well as traditional understandings of the labor process to include the management strategies that also reach beyond the workplace and into the labor market and involve other local institutional actors to help reproduce production and existing social relations.

Table 1.1. Work regimes and commitment compared

Work regimes	Despotic	Panoptic	Peripheral HR	Collectively negotiated
Firm	Allied-Power	Storage Ltd.	Integrated Production	Discrete Manufacturing
Major product	Phones, adaptors, transformers	Hard disk drives components	Integrated semiconductor chips	Discrete semiconductor chips
Technological intensity	Low	High	High	Moderate
Business/ Marketing strategy	Low volume, low cost, standard quality	High volume, low cost, semi-customized	Variable volume, high cost, customized	High volume, moderate cost, semi-customized
Nationality	Korean	Japanese	American	European
Flexibility strategy	Numerical flexibility	Engineer- and surveillance-led flexibility	Engineer- and Management-led flexibility	Engineer-based and Management-initiated flexibility
Work organization	No multi-skilling, no teams, no quality programs, little worker autonomy	Assembly-line, no teams, little worker autonomy	Modular assembly, some larger teams, some autonomy	On- and off-line problem solving, some autonomy through collective bargaining
Localization/ Recruitment strategy	Screening for simple labor market vulnerability	Screening for complex labor market vulnerability	Screening for attitude, education, skills	Internal labor market, referral system

	Low commitment	Coerced commitment	Purchased commitment	Bargained commitment
Production worker profile	Young, single inexperienced women with lower education level	Young, educated, single, inexperienced, women from rural areas	Young, highly educated, inexperienced women	Older women, mixed levels of education
Labor relations	Union suppression and adversarial labor relations	Union prevention through intimidation	Union prevention & substitution through 'soft' HRM	Union acceptance
Nature of worker commitment				
Loyalty	Low	Low Based on occupational prestige	Moderate Based on behavioral training	High but Dual Based on bargained concessions and autonomy
Effort	Moderate Based on direct control	High Based on technical and bureaucratic control	High Based on technical and normative control	Moderate/High Based on "democratic Taylorism"
Attachment	Moderate Based on labor market vulnerability	High Based on over-time pay, labor-market vulnerability	Moderate Based on benefits, low occupational prestige	High Based on bargained job security, pay and benefits

character of their products and product markets, the technological intensity of their production processes, their labor requirements, their employee relations, their wages and benefits, their recruitment strategies and worker profiles, their interaction with local institutions, and finally, the nature of commitment among their workers.

The first case, which I call the despotic work regime, is represented by Allied-Power, two firms from a multinational group of associated companies. Both firms are Korean subsidiaries with a total of approximately twelve hundred workers producing CB radios, electronic transformers, and cellular phone adaptors for other subcontracting Korean firms. The despotic work regime is characterized by the production of low-tech, low-quality products in a traditional, labor-intensive, assembly-line fashion. Management maintains often coercive control over production workers, and flexibility is maintained by either work intensification or labor shedding. The workers receive at or below minimum wages, little training, few benefits, and enjoy little job security. Labor relations at the two plants are highly adversarial. In terms of localization strategies to keep production going, the firms recruit primarily young single women with modest education and little work experience. The firms have also consistently suppressed labor organizing efforts, drawing heavily on local and national state actors to aid in direct and often illegal union-busting. Due to the high levels of coercion both inside and outside the factories, production is unstable and workers show low overall commitment. Thus in terms of the central argument regarding the importance of worker commitment for stability and competitive performance, Allied-Power is the exception that proves the rule.

The second case is the panoptic work regime, found at Storage Ltd., a wholly owned subsidiary of a leading Japanese electronics conglomerate. Storage Ltd.'s $124 million plant, established in 1996, assembles and tests high-end hard disk drives (HDDs) for computer network servers and employs more than nine thousand workers. The panoptic work regime is characterized by quite high-tech yet labor-intensive production. High quality and high functional flexibility is engineer led, supported by a computerized assembly line and electronic surveillance of production workers. The work pace is fast, and managers maintain extremely tight workplace controls through "disciplinary management" and strict enforcement of company rules. But to appease workers, the firm also provides all legally mandated benefits, free transportation, and above-average take-home pay due to high levels of daily forced overtime. The high level of workplace control is coupled with a deep localization strategy. Working through local officials and the private zone manager, the firm recruits well-educated, young, inexperienced, single women from provincial towns and uses these same institutions to maintain surveillance over workers and prevent collective organization. Thus despite onerous conditions at work, collective and individual resistance is muted and worker attachment is relatively strong. By employing a heavily Taylorist work organization and strategically amplifying the conditions of structural inequality through recruitment, Storage Ltd. can be said to construct alienative or "coerced commitment."

The third case, the peripheral human resource work regime,[5] is embodied by Integrated Production, a wholly owned subsidiary of an American semiconductor firm with thirteen hundred employees producing and testing advanced integrated circuits, or IC chips, in a state-of-the-art $200 million plant. Loosely patterned after work organization in similar Silicon Valley firms, this work regime is distinguished by cutting-edge technology and capital-intensive, automated production. Flexibility relies on engineering talent aided by automated information systems that track both workers and production. Management maintains hierarchical control at the point of production, but also uses a range of strategic human resource practices—providing technical and behavioral training, performance-based pay, and above-average benefits, along with good communication—to construct a strong corporate culture. With an emphasis on internal commitment and control strategies, the firm relies less on its localization strategy, though it does recruit inexperienced, but well-educated female production workers; takes strategic action to detect, defuse, and thwart union organizing; and has negotiated with the national government to be exempt from certain national labor laws. Worker satisfaction and loyalty are relatively high, but so is turnover, due mainly to the frustration of educated workers with their low status as operators. The combination of high positive incentives and strategic individualized employee relations are hallmarks of "purchased commitment."

Finally, the fourth case, the collectively negotiated work regime, is represented by Discrete Manufacturing, a branch plant of a European electronics multinational that produces a wide range of discrete (versus integrated) semiconductor chips and employs nearly thirty-five hundred workers. This work regime is differentiated by a medium-level of technology and experiments with new forms of primarily labor-intensive work organization. Flexibility is achieved by engineer and technician-led quality improvement groups. Unlike the other firms, Discrete Manufacturing is unionized, forcing management to bargain collectively with workers over compensation, benefits, and job security in exchange for retaining management prerogative over production. The workforce is experienced, older, with mid-level education, and primarily women who enjoy job security and mobility, well above-average pay, and extensive benefits. Worker commitment is high, but loyalties are of-

---

5. I use the term *peripheral* to denote some common features of the labor market, state regulatory system, and employment relations often found in "institutionally thin" developing countries such as the Philippines.

ten split between the firm and the union. In terms of localization strategies, strong worker bargaining power has tempered and limited management's choices. Unionization has also helped win high wages and job security for the largely female workforce, leading to a relatively stable form of collectively negotiated commitment.

Navigating Suspicion: A Note on Methods

I conducted the field research for this book in the Philippines from January until December 1999 and employed multiple methods to tackle my varied research questions. Not only did I want to document changes in how production is organized in a fiercely competitive industry, but I also wanted to understand *why* change was occurring, and what meanings these changes had for both localities and workers. I therefore chose to use a range of methods, focusing primarily on nonparticipant observation and open-ended interviews, but I also utilized participant observation, semistructured interviews, more formal interviews, and questionnaires. To supplement my findings, I collected primary and secondary written and statistical data on individual firms, the electronics industry, and the national zone program, the various localities and local labor markets, and workers' organization from a variety of sources, including firms themselves, the electronics industry association, published and unpublished local academic studies, various government agencies, local municipalities, labor unions, and nongovernmental community organizations. I conducted several weeks of follow-up research in July 2003. It was, of course, critical for me to place myself within the context of the research process. I went into the field understanding that my own multiple positions, statuses, and identities—as a male, Filipino-American university researcher—would have an impact, not only in terms of my work and its perspectives, but also in the ways that interview respondents would react to me and try to "place" me politically, socially, and culturally during our interactions. This proved particularly important in navigating access to the many different actors in and around the firms and zones.

I began by interviewing the faculty, staff, and especially students of the University of the Philippines, School of Labor and Industrial Relations, as many of them worked in or had ties to the industry, particularly in human resources departments. These early contacts proved invaluable as they provided "proper" introductions to both key government and industry "gatekeepers." Given the extremely tight security apparatus surrounding the export processing zones, just getting into the zones, let alone gaining access to the actual firms or the workers in the firms, can be daunting. Several of the Philippine zones have experienced high levels of labor militancy and industrial action in the past and have also been the target of recent international campaigns by labor rights groups that have publicized and politicized the working and living conditions within the zones (for example, see Klein 2000). This had put key actors on guard. Not only were the firms themselves wary of "researchers," so too was the Philippine Economic Zone Authority (PEZA), the government agency responsible for promoting and regulating the zones.

Getting government permission to visit the zones was thus necessary, but getting introduced by the *right* government agency proved crucial. Several powerful state regulatory bodies have some jurisdiction in the zones, including the Department of Labor and Employment (DOLE) and the Department of Environment and Natural Resources (DENR), but it is PEZA that is often viewed—and views itself—as the most "employer-friendly." Cognizant of the politically charged atmosphere, I contacted the national-level PEZA officials through my affiliations with the university and obtained a list of all electronics firms registered with PEZA and the Bureau of Investments. Following several interviews at the national level, I arranged introductions to and interviews with the PEZA zone administrators on site at one public and six private economic zones. These zones are located in the provinces of Cavite and Laguna, where the majority of foreign firms are concentrated, just south of National Capital Region of Metro Manila.

In large part because the national PEZA office—and not, for example, the Department of Labor and Employment—had introduced me, the local PEZA administrators agreed to share zone-specific data and arrange interviews with electronics firms located within the zones. Stratifying from my master list of firms according to size, nationality, and type of product, I then requested interviews with human resource (HR) managers in four or five firms in each zone. Again, the institutional position of my "sponsor"—this time the local PEZA official—helped opened doors and allay suspicions at the next crucial level—the firm. I initially visited and conducted open-ended interviews primarily with HR representatives at twenty-two firms. I targeted HR managers and staff as key informants at the firm level because the HR manager is almost always the first and sometimes only senior management position that is "localized"—that is, filled by a Filipino. HR personnel are also the most knowledgeable about key commitment and localization strategies, such as pay and promotion, benefits, recruiting, bussing, and connections with local officials and communities.

For selection of the four case studies, I chose to focus on multinational firms of different nationalities in both private and public zones, producing different products, with both unionized and nonunionized workforces. I do not claim that these firms are representative of the entire industry but I do try to capture the range of diversity across the industry. The level of access firms were willing to provide also influenced selection. Because the industry is highly competitive, many firms were quite suspicious of my intentions.

Some firms I requested to visit refused outright, others I did visit would not allow me to even view the production areas, since the layout and arrangement of production flows were considered proprietary secrets, which they guarded quite closely from their nearby competitors. The firms I focus on in this book generously provided extensive access to their facilities and personnel. Except in one case, these firms were in many ways proud of their facilities and human resource policies, and the Filipino HR managers often stressed the benefits their companies were bringing to both employees and to the Philippines more generally. Initially, I developed a semistructured interview schedule to collect basic data on HR practices and work organization. Then, I conducted more open-ended interviews with HR managers, CEOs, supervisors, trainers, quality inspectors, engineers, and other staff. Interestingly, management employees almost always preferred to be interviewed in English, in large part to demonstrate their fluency to a foreign researcher and highlight the status that went with it. As I became more familiar with the plants and to the personnel, I increasingly conducted plant-level observations. I attended technical and behavioral trainings, sat in on employment interviews and tests, traveled with and aided staff on recruiting trips and background investigations into rural areas of two local provinces, and observed production on the shop floors. In one case, Discrete Manufacturing, I joined a project group of on-the-job trainees in the HR department to update the employee handbook. Because I had already conducted interviews at many HR departments, they felt I could help them "benchmark" with other leading electronics firms in the Philippines. Interviews with company personnel, which lasted from one to two hours, were generally not recorded, in part to encourage candid responses. Following interviews and in-plant observations, I tried to write up my notes immediately, often while sitting in my car in the company parking lot. In all, I conducted thirty-six formal interviews with management personnel within the firms in addition to plant-level observation, participating in trainings and more informal interactions during lunch breaks and between shifts. While initially I hoped to conduct participant observation actually working on the production floors, the training requirements and the extremely strict rules governing the International Standards Organization (ISO) 9002 quality certification for all workers in three of the four case studies made such arrangements impossible. I thus chose to rely primarily on open-ended interviews and nonparticipant observation.

While gaining access to the firms and production floors was somewhat difficult, being able to interview workers at the workplace proved even harder. The same set of "gatekeepers" and sponsors that helped me win the trust of management were unfortunately the same set of actors that made workers suspicious of both my affiliation and my intentions. Inside the plants, a company representative nearly always shadowed me. Even when I was observing work on the shop floor, the shift supervisor generally introduced me to workers and the color of my "bunny suit" often tagged me as a supervisor. Despite multiple explanations of my status as an independent researcher, workers were understandably cautious. In some cases, the level of surveillance and intimidation were palpable. For example, while I was conducting observations at Storage Ltd., a senior production engineer explained the production process and encouraged me to ask questions of any employee. Yet as we stopped at one station, she explained that the current worker who did not look up during the supervisor's comments, was one of her best and was invaluable—so invaluable, in fact, that she had no replacement and thus worked seven days a week. When I tried to ask the operator directly what happens when she gets sick, the production engineer quickly replied—as much to me as to the operator herself—"Oh no, she doesn't get sick."

At the other firms, I did not sense that same level of intimidation. Nevertheless, given the sensitive and critical nature of my questions and since many workers in the plants considered me directly aligned with management, I did not feel that workers would candidly discuss with me issues related to control and commitment. Thus interviews with workers, for all cases expect Discrete Manufacturing, were arranged and conducted outside the plants without company participation. At Discrete Manufacturing, the union helped in identifying potential respondents and allowed me to interview workers in a union-designated area inside the cafeteria. For interviews conducted outside the plant, I again found it crucial to approach my informants through trusted gatekeepers.

In an early attempt to contact Storage Ltd. workers, I first tried going directly to boarding houses around one of the firms' shuttle bus pick-up points where workers were encouraged to relocate. While I did identify a boarding house occupied primarily by Storage Ltd. workers, they were extremely reluctant to grant me an interview. At first, I thought this reluctance sprang from a fear that I might be affiliated with a labor union. Yet as it turns out, these workers were in fact far more afraid that I was representative from the company sent to check up on them and test their allegiance to the firm. From then on, I tried to approach workers through people they already knew or trusted. For example, the parish priests in two localities proved essential when approaching boarding houses and their owners, as no other males were allowed to visit the sex-segregated boarding houses, particularly in the evenings after work when I conducted many interviews. At other times, I relied on referrals from previous interviewees (snowball sampling), other church leaders, boarding house owners, other respected community members, and local officials.

An open-ended interview schedule was developed with the help of local researchers from the Workers' Assistance Center and the Ecumenical Institute for Labor Education and Research. Interviews were conducted in Filipino by the author and several research assistants. Several local female researchers were especially helpful in conducting the interviews, in large part because it was easier for them to gain access to the boarding houses and often women workers felt more comfortable and were more candid when speaking with a local woman close in age to themselves. In all, seventy-five interviews lasting from one to two hours were conducted with workers in their homes or boarding houses. Twenty-nine of these interviews were tape recorded, translated, and transcribed. For the remaining interviews, extensive written notes were taken, both during and immediately following the interview.

Finally, I also conducted interviews and participant observation in local communities and with several church and labor organizations. I attended church services and informal after-church activities, and conducted community visits with priests and seminarians based in the communities. I also attended worker education seminars and informal get-togethers at a workers' community center, and attended several union meetings and rallies both within and outside the zones. Community-members interviewed included boarding house owners, priests and parishioners, mayors, lower-level officials (*barangay* captains), members of the Industrial Police Force, provincial employment office representatives, and workers' families.

Overview

In chapter 2, I set the stage for analyzing the making and impact of high-tech production by examining three major macro-level influences on how work is organized and negotiated. These include the conditions and demands in the global electronics industry, the Philippine state's development and foreign investment strategies, and the general characteristics of the Philippine labor market.

In chapter 3, I bring the analysis down to the factory level, examining in detail the four work regimes and their diverse strategies to organize work and employment relations and their impacts on workers. Specifically, I compare production technology; flexibility, quality and productivity strategies; policies of labor control and employee relations; organizational hierarchies; and finally, workers' collective versus individualized negotiating power.

I extend the analysis of work regimes in chapter 4, highlighting the broader political and social context within which work is negotiated. Here, I discuss the firms' strategic localization practices, or how their concerted labor control and commitment strategies stretch far beyond the factory and into the local labor market. Alongside this discussion, I also profile the evolution of the Philippine export processing zone (EPZ) program, as its development is intimately connected to changes in investment and work organization.

In chapter 5, I turn to the workers themselves, drawing on extensive interviews to trace their subjectivities and the meanings they attach to work. I focus specifically on their perceptions of employment conditions, points of reference and comparison, and their own definitions of "autonomy" and job satisfaction.

Finally, in the concluding chapter, I sum up and knit together the book's major themes: the changes in technology and competitive demands, the reorganization and localization of production, and the manufacturing of worker commitment. I then assess their wider meanings, both for applicability in other contexts, and for our broader understanding of how globalization processes are locally constituted.

2 Global Electronics, Filipino Workers, and the Regulatory State

> Meet Diego, one of the 300,000 Filipino workers who make state-of-the-art technology products. Highly skilled in his craft, he can work at world-class productivity levels and come up with a zero-defect product. He practices the best-known methods in manufacturing and his extensive capability ranges from the assembly of the smallest components to entire finished products. . . . Diego makes high quality, globally-competitive electronics products of American, German, Korean, Singaporean, Taiwanese, Dutch, Swiss, Swedish, Malaysian, English, Scottish and Filipino electronics companies. . . .
>
> Every single Filipino worker in these companies are [sic] truly world-class. In a mere 6–8 weeks, their technical skills are already honed. A far shorter learning curve when compared with their Asian counterparts. They are not only multi-skilled but proficient in English as well.
>
> The success of the Philippine electronics industry is powerful proof that the Filipino worker is a cut above the rest.
>
> The Great Filipino Electronics Worker
> *Panlaban Sa Mundo* (World Class)
> Philippine Electronics Industry
> Promotional Campaign (SEIPI 2003)

GLOBALIZATION presents an interesting paradox: the less place-bound capital becomes, the more important places actually are. For manufacturers, greater mobility, the revolution in communication technologies, and a more liberalized world economy has meant the dogged competitive quest to locate particular slices of their production chains in places that offer the best resources to do the job. Localities and states hoping to draw such investments find themselves caught in competitive tournaments, trying to promote their unique local assets and conditions that are not found on the same terms elsewhere. At the intersection of both firm and state strategies then is the availability and reproducibility of particular, place-bound factors of production.

As the brochure quoted above illustrates, for the global electronics industry and the Philippines, the crucial place-bound asset is high-quality labor, available at the right price and under the right conditions.

This chapter examines why the global electronics industry has come to the Philippines and how it organizes production. First, I survey the global electronics industry and how the changing complexity of accumulation has pushed high-tech firms offshore, but beyond a simple hunt for cheap female labor. Second, I explore how the Philippine state has attempted to lure such foreign investment as a centerpiece of its revamped—if misguided—development strategy. I focus specifically on the state's divergence from an East Asian developmental state model since the 1970s and its crystallization in the 1990s as a neoliberal regulatory state, promoting "glocal developmentalism" or global competitiveness within strategic sub-national sites (Brenner 1999). In this case, the Philippine regulatory state provides investors with high-grade infrastructure and stable local conditions in its upgraded export processing zones (EPZs) for unmitigated access to the country's primary competitive advantage: inexpensive, English-speaking workers. The final section of the chapter profiles the unique "fixities" of this key resource—labor—and the social relations and gendering processes surrounding its sale, purchase, and reproduction.

Global Electronics

Changing Character of Production and Competition

There is no doubt that electronics is a key manufacturing sector for the twenty-first century, especially the semiconductor and computer hardware subsectors. As the "industry of industries," semiconductors have had a broad transformative impact because of their applicability through computerization to all major sectors of the economy, from automobiles, defense, and telecommunications to banking, finance, and agriculture (Dicken 2003). Electronics has also been a bellwether of globalization, being one of the first sectors to move "offshore" and disperse its manufacturing globally in the 1960s. And the electronics industry has been a pioneer in developing new digital communication and production technologies toward organizing more complex cross-border production networks in the face of intensified global competition since the 1990s.

For developing countries, electronics represents one of the very few sectors that offer even low-level producers *potential* access to higher technologies, knowledge-intensive industries, higher wage employment, and participation in innovative global production networks (Borrus, Ernst, and Haggard 2000). The oft-cited East Asian "success stories" of South Korea, Taiwan, Hong Kong, and Singapore to a large extent built on their initial participation in global electronics to jump-start their advanced industrial-ization via export-led growth and industrial upgrading. Yet despite being central to past successes, these export-led models may not be able to offer sustained growth and broad-based economic development in the contem-porary era of both increased economic liberalization and more complex competitiveness.

Electronics Overview

To explore the potential of the electronics sector for developing countries like the Philippines, it is important to have an overview of the sector, un-derstand the character of the value chain across different segments, and un-ravel how the changing nature of innovation and manufacturing have shaped current locational decisions. The $1 trillion a year electronics sector covers a wide array of products but the key manufacturing industries for de-veloping countries have been primarily semiconductors, computers, and consumer electronics. As we shall see below in the breakdown of the value chains and locations of these subsectors, all have been dramatically shaped by the speed and intensity of technological change and the ferocity of mar-ket competition.

The semiconductor industry is the backbone of the entire sector, provid-ing the key components for a vast array of other products. In the 1950s, the first integrated circuit (IC) was developed, allowing for the connection of a large number of transistors etched on a single tiny "chip" of silicon, the size of a fingernail. Driven by the increasing demand for sophisticated micro-electronics, the semiconductor industry has been one of the fastest growing industries in the world, exploding from a US$100 million industry in the 1950s to a US$200 billion industry in 2000. Although the industry was hit hard by the bursting of the internet bubble and the slowdown in the sales of personal computers in 2001, the industry grew 10 percent in 2003 and again accelerated to a 27 percent growth rate in 2004 (SIA 2005).

The semiconductor subsector can be broken down into different kinds of semiconductors, with different market demands and characteristics. The first divide is between discrete semiconductors and integrated circuits. Discrete semiconductors, used primarily in telecommunications and consumer prod-ucts, are somewhat less flexible and less technologically sophisticated than IC "chips." Logic chips, which make up the largest share of the IC market, include innovative and capital-intensive technologies, including micropro-cessors—the brains of a computer—and digital signal processors, used in communications, consumer electronics, and computers (Linden et al. 2001). An equally important element in the computer industry boom has been the development of the hard disk drive (HDD), which provides inexpensive mass storage of retrievable electronic data. During the 1990s, the industry grew more than 20 percent per year, and by 1999 shipped more than 150 million

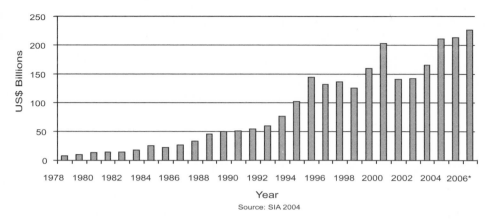

Figure 2.1 Worldwide semiconductor history and forecast

units worldwide and had annual sales of more than $30 billion (Disk/Trend 1999).

Both the semiconductor and HDD industries are characterized by technological dynamism and rapid change that pose significant challenges for manufacturing. The pace of innovation in the semiconductor industry (and therefore the shortness of product life cycles) is legendary. Intel founder Gordon Moore boldly predicted in 1965 that the number of transistors per chip would double every one and a half years. Indeed, Intel's first microprocessor chip, developed in 1971, was a technological breakthrough consisting of 2,250 transistors. By 2000, the company's Pentium 4 microprocessor chip— packaged and assembled in the Philippines—contained an astounding 42 million transistors (Dicken 2003). HDD technologies have experienced similarly spectacular leaps and bounds. For example, the areal density, or the amount of information that can be stored on a square inch of disk, grew at an impressive 30 percent per year until 1991. In the 1990s, the rate of technological change actually began increasing, reaching 60 percent per year between 1992 and 1997 and a staggering 125 percent in 1999, a rate even *faster* than Moore's prediction for semiconductors (McKendrick et al. 2000).

Value Chains—Production Sequence

For both industries, maintaining such rapid technological innovation requires increasingly high capital expenditures for research and design, extremely sophisticated "clean" manufacturing facilities, and tightly integrated—if globally dispersed—supply chain management. While the production processes for semiconductors and HDDs differ in important ways, as high-technology industries they share fundamental value chain segments, which include design, development, and engineering; wafer and/or key com-

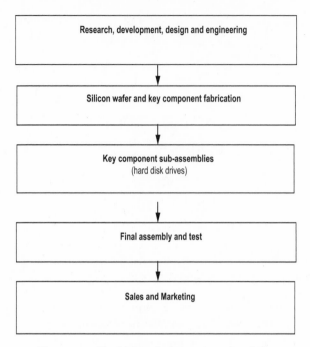

Figure 2.2 The high-tech electronics value chain

ponent fabrication; subassemblies (for HDDs); and final assembly and testing. Each of these stages or processes in the production sequence has different technological, manufacturing, and labor demands, which greatly influence their distribution processes across space.

RESEARCH, DEVELOPMENT, AND DESIGN ENGINEERING   As both industries are driven by continuous technological innovation, it is no surprise that research and development are at the heart of leading companies' "core competencies." Intel, for example, has been able to maintain its impressive lead in microprocessors in large part because of its proprietary technology, design architecture, and its heavy investment in successive generations of increasingly more powerful chips (Spar 1998). Chip design and process development require pools of highly skilled scientists, engineers, and technicians, often working in strategic alliances with universities and other firms. The need for such knowledge intensity has kept much of such work agglomerated in high-tech clusters near corporate headquarters in the United States, Japan, Europe, and to a growing extent, South Korea and Taiwan. In the 1990s, when U.S. firms reasserted their dominance in higher-end semiconductors, design, prototyping, and production needed were kept primarily in California's "Silicon Valley" in order to maximize the speed at which new products could be developed and ramped up for high-volume production (Angel 1994).

WAFER AND KEY COMPONENT FABRICATION   The core manufacturing process for high-tech industries such as semiconductors and HDDs is wafer fabrication, or the actual etching of the circuits of individual "chips" (up to five hundred per wafer) onto successive layers of silicon. Wafer fabrication is extremely technical and competition is driven in large part by firms that can achieve the highest yield, that is, the fewest rejected chips per wafer. Fabrication thus requires the same level of engineering talent as in the design phase and increasingly expensive manufacturing plants. Today, a new wafer fabrication plant can cost $3 billion, making it one of the most capital- and research-intensive manufacturing industries in the world. The high capital and quality requirements, the need for intense interaction with chip designers, the demands for top-notch engineers, and need to ramp up production quickly kept the bulk of wafer fabrication clustered near research and design centers, at least through the mid 1990s. Increasingly, wafer fabrication has spread to other areas with the rise of Taiwanese and Singaporean foundry firms that manufacture chips according to customer design specifications in different parts of East Asia, especially China (Ernst 2003; Linden et al. 2001).

KEY COMPONENT SUBASSEMBLIES   This portion of the value chain is more specific to HDDs and other complex electronics that require precision moving parts and combinations of key components. For HDDs, these subassemblies include manufacturing tiny precision motors and read/write heads or sliders, which are then "stacked" to increase storage capacity. Semiconductor chips that help operate the drive must also be mounted or "stuffed" onto printed circuit boards that are then inserted into the drive. These subassemblies, as they have miniaturized and become more sensitive to environmental contaminants, are produced in expensive "clean" environments and have become more automated. Yet some processes still remain labor intensive.

FINAL ASSEMBLY AND TESTING   Both semiconductors and HDDs share a similar break in the production process at the final assembly and testing stage that sets it off from the more capital- and engineering-intensive design and fabrication stages. For semiconductors, final assembly is the stage in which the individual chips on the larger silicon wafer are ground thinner, individually tested, broken off, mounted onto lead frames, connected to the leads using extremely thin gold wires, then encapsulated or "packaged" within plastic for protection. The complete chips are then "burned-in" or individually tested under extreme conditions for reliability during normal use. For sophisticated IC chips, all final assembly must also take place in a clean room environment. For less sophisticated discrete semiconductors, some portions of assembly do not require clean rooms. For HDDs, finally assembly is even more complicated. It brings together the subassemblies of motors, printed circuit boards, media or disks, and the head stack assemblies. These parts are then mounted and all parts are enclosed into a sealed aluminum casing. Finally, the HDDs go through rigorous electronic and functional testing.

For both semiconductors and HDDs, final assembly and testing remain quite distinct from design and fabrication, have historically been the most labor intensive portions of production, and thus the most likely to be relocated to areas around the world with low labor costs. In fact, the search for low assembly labor costs are what first led lead firms from the United States, Japan, and Europe to invest in manufacturing facilities abroad, particularly in East and Southeast Asia beginning in the 1960s. Today, production has become more automated and usually requires clean room environments. While assembly plants are *relatively* cheaper than wafer fabrication plants, stricter production standards and higher technology requirements have pushed up the price of assembly and testing facilities, which now cost from $100 million to $500 million dollars. Nevertheless, labor remains a key factor: IC chip assembly and testing plants still usually employ between fifteen hundred and four thousand workers, while HDD assembly plants can employ upward of nine thousand mainly low-skilled and semiskilled workers. Wages are thus the most important variable cost, accounting for 25 to 30 percent of total operating costs at the assembly and testing stage (Spar 1998). But the assembly and testing of high-tech products like ICs and HDDs also require a bevy of skilled technicians and process engineers to ramp up, adjust, upgrade, and maintain volume production. Thus the availability of relatively inexpensive technical and engineering labor is also a requirement at this stage of production.

Finally, an ongoing tension at the assembly and testing stage is the optimal level of automation. While automation can improve quality and yields, specialized equipment is also expensive, depreciates quickly (due to short product cycles), requires longer production runs to recoup costs, and is thus less flexible and cost effective for rapid production ramp ups or ramp downs. Manual assembly, which produces somewhat lower yields, is a more flexible and variable cost that can more easily be "brought on line" or "retired" in response to demand or production swings.

Market Competition in Advanced Electronics

Adding to the technical pressures in production is the both fierce and complex character of market competition in semiconductors and HDDs. Price wars for end products like PCs and consumer electronics are now endemic, which has meant a shift for chip and disk drive producers from relatively high profits in the 1980s to increasingly low or razor-thin profit margins for even the most sophisticated products within six months of their introduction. The hard drive sector, for example, is now considered a "commodity industry" because of its standardized, undifferentiated product and its intense competition based primarily on price per gigabyte (Bohn 2000). Stiff competition, higher capital intensity, and high research and development costs have all contributed to industry consolidation. From the already small number of fifty-nine companies that produced disk drives in the early 1990s, only eighteen were active in 1999 and only eight in 2004 (Chellam 2004; Disk/Trend 1999). American companies traditionally dominated the industry, but Japanese firms began to challenge American dominance in the 1990s. By the late 1990s, market share for the four major Japanese companies increased to 30 percent. It is interesting to note here that all four major Japanese HDD players have set up wholly owned subsidiaries in the Philippines since 1994 and three have chosen the Philippines as their exclusive site for HDD assembly (Tecson 1999).

In semiconductors, the most intense price competition has been in memory chips, but even in other areas, such as microprocessors, price declines in final goods have also led to commodity-type markets. By 2004, the $213 billion semiconductor market was still led by data processing or computing, but the rise of wireless communications and consumer products has boosted their share of chips significantly (SIA 2005). The lack of a single, standard-controlling firm in the data and telecommunications segment has only led to greater competition among leading firms such as Samsung, Motorola, Texas Instruments, Philips, STMicroelectronics, and NEC.

These "high-tech commodity" industries are thus increasingly dominated by a handful of leading global firms competing on the basis of cost, quality, product differentiation, *and* speed-to-volume. Dubbed "mass customization," the main industry strategy combines the characteristics of mass production and economies of scale with extremely short product cycles, niche markets, and rapid technological innovation. Competitive success depends on a capacity to orchestrate complex international production networks to bring innovative products to market quickly, in order to reap first-mover pricing advantages. These production networks, which are increasingly stretched across the globe, consist of a lead firm, its subsidiaries, affiliates and joint ventures, its suppliers and subcontractors, and its distribution channels and alliances (Ernst 2002).

Internationalization of Electronics Manufacturing

One of the first responses by leading electronics firms to growing competition and technological change was to reduce costs by dividing their value chains and shifting production abroad. The advantage of moving some production offshore was facilitated by the liberalization of the international trading and investment regimes from the 1960s onward as well as improved communications and transport technologies. While competition has become increasingly complex, there remains a fundamental break in the value chain and the strategic location of key production clusters. The capital- and engineering-intense clusters of R&D, design, and wafer fabrication, what Ernst (2002) has called "centers of excellence," focus on raising competitiveness through innovation and improving yields, and have remained quite concentrated. On the other hand, labor-intensive assembly and testing plants are what Ernst refers to as "cost and time reduction centers." The emphasis on increasing flexibility and efficiency while lowering overall costs has meant that assembly and testing has been far more globally dispersed.

The internationalization of the electronics industry began in the 1960s with IBM's decision to shift its assembly to Japan and then to Taiwan. Despite high setup costs, the low wages in Taiwan made hand assembly cheaper than full factory automation at home. And by hiring women, whose labor is systematically undervalued, IBM could reduce its labor costs still further. The success of IBM and the logic of employing low-paid Asian women became a model for the industry. In 1968, Texas Instruments and National Semiconductor moved to Singapore and then both set up IC assembly plants in Malaysia in the early 1970s. At about the same time, Intel and Texas Instruments also opened assembly plants in the Philippines (Ernst 1997; Aldana 1989).

To contain "technology leakage" and maintain equity control, the semiconductor firms chose to invest directly into 100-percent-owned affiliates. The production was strictly labor-intensive, screwdriver assembly with very limited local value-added and few local linkages. Offshore sites provided a low-cost assembly and export platform, while all product and process design, subcomponent manufacturing, and marketing remained at home (Henderson 1989).

The appreciation of the U.S. dollar in the early 1980s and the Japanese yen in 1986 provided new impetus for multinational corporations (MNCs) to relocate an increasing amount of their production abroad. This entailed the upgrading and automation of foreign affiliates in Asia and increased need to work with more local suppliers in the region. Particularly in consumer electronics, as local subcontractors gained increasing experience in manufacturing and developed more sophisticated original equipment manufacturing (OEM), there was also a shift away from wholly owned affiliates toward the use of interfirm production networks. These developments paved the way for early subcontractors in South Korea, Taiwan, and Singapore to move up the value-added chain from simply contract assembly to more mature OEM or contract manufacturing.

The 1990s witnessed an explosive growth of the semiconductor and electronics industries around to world, but particularly in Asia. The Asia/Pacific has been the fastest-growing region in the world, growing from less than 15 percent of revenues in 1990 to more than 41 percent by 2004 (SIA 2005; Austria 1999). Within the Asia/Pacific, a growing regional division of labor has emerged. Japan has for a long time been the Asian leader in electronics manufacturing and remains the second largest producer of electronics after the United States. Like the leading American firms, Japanese MNCs provided the manufacturing design and technology, which they develop at home but implement throughout the region. However, South Korea and Taiwan have continued to upgrade their own industries since the 1980s, increasingly adding design and development to their already sophisticated wafer fabrication capabilities. By the late 1990s, South Korea had become the world's third largest electronics producer. The relatively advanced countries of Southeast Asia—Singapore, Malaysia, and Thailand—have become the hosts of some advanced and standardized products since wage rates rose in South Korea and Taiwan. Finally, labor abundant countries such as China, India, Indonesia, and Vietnam have become the preferred sites for labor-intensive assembly and testing. The Philippines currently straddles the two lowest categories, with most of its production in labor-intensive assembly but increasingly with higher, more capital-intensive products such as Texas Instruments' digital signal processors and the latest generation of microprocessors and flash memory from Intel.

However, the biggest development in the Asia region has been the meteoric rise of China, which has since 2000 become the undisputed leading site for a wide range of production, from fabrication to assembly and testing, and increasingly, for design and engineering. In 2002, China passed both Taiwan and South Korea to become the world's third largest producer and as well as the third largest market of electronics, capturing nearly 9 percent of both the world's output and total market (*Yearbook of World Electronics Data* 2004).

Locational Strategies

The rise of China demonstrates that while electronics production has become dispersed, it is still concentrated in a relatively small number of Asian countries. So what are the main determining factors for where high-technology production gets (re)located? Low wages, while still important, are by no means the only factor. For example, by moving HDD assemblies initially from Silicon Valley to Singapore, firms in the 1980s were able to reduce the share of labor costs in assembly from 25 percent to 5 percent (McKendrick et al. 2000). By the 1990s, the share of labor costs to total costs was down to 3 percent (Wong 2001). With extremely low profit margins, shaving even a small amount off labor costs is still a priority. However, with the rising capital intensity and the changing character of competition, cost and competitive demands have become more diverse.

As theorists of foreign direct investment have argued, transnational firms looking to reduce costs do not simply look for areas with the lowest wages, but seek out "location-specific factors," or assets not widely available elsewhere (Dunning 1993). The key location-specific factors for advanced electronics production include available infrastructure, a clustering of other industry firms, labor cost and availability, political stability, and the regulatory environment (Dicken 2003; McKendrick et al. 2000; McMillan, Pandolfi, and Salinger 1999).

Infrastructure

As electronics production has become increasingly technology- and capital-intensive even at the assembly and testing phase, the need for generic and industry-specific manufacturing infrastructure has increased. In particular, semiconductor production itself requires reliable and uninterrupted power supplies, vast quantities of pure water, and wastewater treatment facilities. In addition, as speed-to-volume production has become a key competitive driver, efficient road networks and easy access to airports and other transportation infrastructure have become essential. Extremely short product cycles and the demands of customization have also heightened the need for high-quality communications infrastructure, in particular high-speed voice, video, and electronic data networks. Finally, efficient customs and import/export procedures are also important for bringing products quickly to market (McMillan, Pandolfi, and Salinger 1999).

Those locations that have successfully drawn advanced electronics investments have often had or were willing to develop such infrastructure. For example, in the early 1980s, Singapore successfully drew the initial foreign investments from the U.S. HDD industry in part because it was already far ahead of other Southeast Asian countries in terms of electricity production and reliability, and the number of telephone lines per capita (McKendrick et al. 2000). Other, less-developed countries in the region hoping to draw investment tried to emulate Singapore. But as we shall see below in the case of the Philippines, the provision of such high-quality infrastructure is expensive, takes time to develop, and is extremely expensive to deliver on a nationwide scale. Given limits on resources and time, "follower" states like the Philippines have used a much more circumscribed infrastructure development strategy, aimed at the provision of high-quality, but privatized infrastructure, restricted to designated export processing zones (EPZs).

Agglomeration Economies

Economic geographers have long noted that businesses tend to cluster in particular places, in large part because proximity to other businesses, customers, or specific labor pools can create positive spillovers that can reduce transaction costs and uncertainties and/or spur innovation (Dicken 2003; Piore and Sabel 1984). In the electronics sector, a divergence in locational clusters has closely paralleled the main divide in the production chain between "centers of excellence," which focus on technological innovation, and "cost and time reduction centers," which focus on making production more efficient. Santa Clara, California, known to the world as Silicon Valley, is often held up as the paradigmatic, innovation-centered "new industrial district" for its concentration of leading electronics firms such as Fairchild Semiconductors, Intel, Cisco Systems, Apple Computers, Sun Microsystems, and Hewlett-Packard (Saxenian 1994). The success of Silicon Valley and the industry's continued "stickiness" or concentration despite rising costs is due to the needs for deep pools of specialized labor that facilitate research and technological diffusion, and for the face-to-face interaction between venture capital, designers, producers, and suppliers to move from prototyping to production quickly (Markusen 1996; Angel 1994).

Cost and time reduction centers focused on assembly and testing manufacturing also benefit from clustering, but have a distinct character. Here, agglomeration is driven not by the need for innovation, but by the market demands for cheaper, faster, and higher quality production. As McKendrick and his coauthors note, the main externalities from HDD assembly clusters in Singapore, and to a lesser extent, in Thailand and Malaysia, are created by the presence of multiple supplier firms, and from specialized labor markets in which managerial, engineering, and operator labor with industry-specific skills are widely available (McKendrick et al. 2000). However, there is little dependence on *local* suppliers, since it has become increasingly common for large lead firms to coax their foreign suppliers to migrate with them when they move production (Tecson 1999). In the Philippines, such imported clusters are created within the EPZs, which are dominated by foreign lead firms and their suppliers. The remaining key "locally specific" assets that a host country can offer, then, are their labor markets and their regulatory policies.

Labor and Labor Market Conditions

The technological and competitive changes in the industry have led some to discount the role of labor in investment location decisions. But while other factors of production—such as machinery and raw material inputs—become less place-bound, labor remains more "sticky" and labor costs more geographically uneven. Particularly in the assembly and testing segment, labor remains a key variable, making up from 25 to 50 percent of operating costs (Spar 1998; McMillan, Pandolfi, and Salinger 1999). Labor is also, of course, not simply another factor of production, but a "fictive commodity" because the labor process and labor markets are inherently social; infused with unequal power relations between workers and owners, and rife with stratifications among workers themselves (Peck 1996; Storper and Walker 1983). For example, despite all the technological changes, women still dominate assembly work, demonstrating the tenacity of gender ideologies that continue to equate low-paid production jobs with women workers (Mills 2003). Because of their stickiness, uneven costs, and ability to dramatically affect the labor process, labor is still often considered "the single most important location-specific factor" in determining investments (Dicken 2003, 210; see also Storper and Walker 1989).

In considering locations, investors look at a range of labor characteristics, not just wages. In particular, they consider the availability and costs of a wide range of skilled and unskilled workers, education levels, industry-specific talent, labor productivity, and the degree of labor militancy. A survey of thirty-nine leading American electronics firms concerning investments in developing countries found that infrastructure, the labor force, and political stability were rated nearly equal as the top considerations for investment entry (McMillan, Pandolfi, and Salinger 1999). Interestingly, when asked specifically about labor force considerations, the firms cited skilled labor cost and availability as their top concern, but this was followed closely by "history of labor unrest or general strikes," "power of unionized labor," and "labor force turnover/stability," all deemed "very important" entry criteria.

The electronics industry is in fact well-known for its vociferous antiunion actions and advocacy, both at home and in their widely dispersed branch plants (Chun 2001; Colclough and Tolbert 1992; Milkman 1991; Hossfeld 1995; McLoughlin 1996; Hodson 1988). High-tech commodity producers, then, focus their labor concerns on cost, availability, quality, *and* controllability. All of which are necessarily influenced by government policies and regulations.

Public Policy, Political Stability, and Regulatory Environment

While it is often argued that globalization and increased market liberalization since the 1980s have weakened or even marginalized the role of national governments, states still play a fundamental role in regulating the economy. Above all, nation states provide the key framework, or "rules of the game," for basic market-regulating institutions such as the financial and legal systems for corporate governance, industrial relations system, education and training system, and the systems of intercompany relations (Soskice 1999; Hall and Soskice 2001). States also continue to provide crucial public goods, such as political and social stability, and the majority of physical infrastructures such as roads, airports, and telecommunications. But despite—or as some have argued, because of—the trends toward market liberalization and "privatization tournaments," states that compete for foreign investment must go beyond simple deregulation. Adopting basic liberalization policies no longer differentiates one state from another. So host governments must make themselves more distinct through deeper "proactive" interventions and the development of "created assets" like a well-educated and trained work force, and industrial and investment policies that cater to specific industry needs (O'Riain 2004; Ernst 2002).

In the electronics sector, the key public policies concern political and economic stability, high-quality labor and labor market discipline, high-quality physical infrastructure, few operational restrictions (such as local content laws), and investment/tax incentives (Ernst 2002). In their study of the HDD industry in Southeast Asia, McKendrick et al. (2000, 232) found that while "the first and most basic prerequisite for entry was the cost of labor," they also note that "labor availability, the flexibility of labor markets, and managerial freedom over shop floor organization (all strongly affected by public policies) were also important variables in the firms' calculus."

A telling example is Intel's selection of Costa Rica as the site of its new chip assembly plant in 1996. The company was drawn first by the political and social stability of this "Switzerland of the Americas" and its aggressive liberalization of the economy. But what set Costa Rica above other competing states was its willingness to offer a broad(er) range of incentives and infrastructures tailored to Intel's needs. These included the company's own free trade zone, an initial twelve-year tax holiday, exemption from import duties, and unrestricted capital repatriation. The Costa Rican government also offered to repay Intel a percentage of its payroll and agreed to subsidize a training program in which the government would pay workers for three months of on-site training, resulting in, "3 months of free labor for corporations in the free trade zone" (Spar 1998, 27). To "close the deal," the government helped engineer a press blackout of the negotiations, the intercession of the attorney general to amend certain unfavorable tax laws, reduced energy rates, improved roads and air traffic scheduling, and redesigned the national curriculum of its technical high schools and advanced vocational training programs to meet the specific needs of the company for trained technicians (Spar 1998, 20–21).

In effect, while policies ushering in market liberalization are necessary, they are not enough. Instead, *selective deregulation* is increasingly being augmented with more sophisticated industrial and investment policies that amount to a *reregulation* of the economy and the labor market that can provide a secure, "probusiness" environment in which firms have more control over the processes and factors of production. This notion of reregulation will be taken up in more detail in the section on the Philippine regulatory state.

The Electronics Industry in the Philippines

> The spatial division of labor is no longer a sectoral one. . . . [In] the future, places will not be known . . . by what they produce as much as who is employed there at certain stages of specific production processes.
> —CLARK et al., cited in Peck 1996, 14.

Beginning in the early 1970s when many electronics firms began shifting assembly manufacturing abroad, the Philippines was seen as an ideal production site because of its large pool of cheap yet educated, English-speaking women workers. While electronics manufacturing has become more technologically advanced, the range, availability, and quality of Philippine labor continues to be the key draw for foreign investors. The Philippines first appeared on the global industry's map in 1974, when Intel and Texas Instruments both set up assembly plants (Ernst 1997; Aldana 1989). By the early 1980s, the Philippines had established itself as a leading export platform and electronics began to play a major role in Philippine exports. In 1976, electronics exports were just US$115 million, or 3 percent of total Philippine exports. By 1982, exports had reached $1 billion, a tenfold increase in just six years, and had already become the country's top exporting sector (DTI figures cited in Aldana 1989).

But the 1980s were a decade of tremendous political and economic instability in the Philippines, with state violence and civil unrest rising as the population chafed under the draconian rule and martial law of then-president Ferdinand Marcos. Although an early leader in attracting investments in East Asia, the Philippines slipped behind its neighbors as both Japanese and American electronics firms opted to invest in other locations such as Taiwan, South Korea, Singapore, and Malaysia. The decline is evident in the stagnation of Philippine exports in electronics over the 1980s: after growing 1000 percent from 1974 to 1982, exports leveled off (and for a period, contracted), growing only a total of 20 percent from 1982 to 1988 (see figure 2.3).

In the 1990s, the growing economic and political stability under President Fidel Ramos helped improve the attractiveness of the Philippines to investors just as the global demand for semiconductors for computers, telecommunications, and consumer products began to expand. Thus by the mid 1990s, the Philippines reemerged as a premier site for advanced assembly and testing. The investments that began pouring in were dominated by leading American and Japanese multinationals such as Intel, Texas Instruments, Cypress Semiconductors, Amkor/Anam, Fujitsu Computers, Hitachi, Toshiba, and TDK. But Korean, Taiwanese, and Dutch investments were also significant. Investments rose from US$44 million in 1992 to US$1.47 billion in 1997. While investments dipped in 1998 to US$671 million, they then rebounded, with investments again topping out at $1.24 billion in 2000 (see figure 2.4). However, inward investments were severely hurt by the global slowdown in the industry and the increasing shift to China. In 2001, new investments were $720 million and fell even further to $270 million in 2002.

Exports

Exports have also grown phenomenally since the early 1990s. In 1990, electronics exports reached US$1.6 billion, or about 19.5 percent of all exports. By 2000, electronics exports climbed to US$27.16 billion and increased its dominant share of Philippine exports to 71 percent. Semiconductors alone

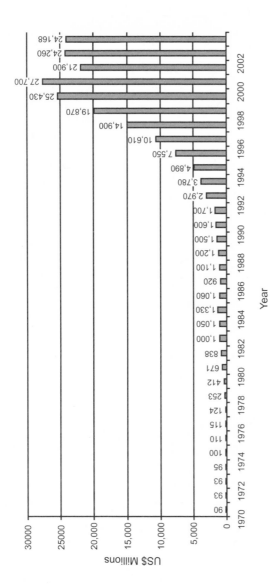

Figure 2.3 Philippine electronics exports, 1970–2003

Source: DTI/BETP cited in SEIPI 2003

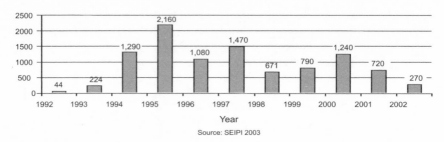

Figure 2.4 Investments in electronics, 1992–2002

accounted for more than 50 percent of the country's total exports. During this period, growth in electronics exports averaged more than 30 percent each year (see table 2.1). The Philippine electronics industry was hit hard by the global slump in electronics in 2001, but has shown signs of rebounding since 2002, with exports reaching nearly $27 billion in 2004.

Employment

The booming global demand for the types of electronics that are assembled in the Philippines has made the industry a leading employer. Employment grew from 69,000, or roughly 5 percent of total manufacturing employment in 1990, to 346,000, or roughly 10 percent of total manufacturing jobs in 2003 (BLES 1999; SEIPI 2003).

Based on the foregoing scenario, it would seem that electronics has been

Table 2.1. Philippine electronics exports, growth, and percentage of total exports

Year	Exports (in US$ billions)	Growth rate	Percent of total Philippine exports
1992	2.97	20	28
1993	3.78	27	33
1994	4.89	28	36
1995	7.55	55	43
1996	10.61	40	52
1997	14.98	41	58
1998	19.87	33	67
1999	25.34	27	72
2000	27.16	7	71
2001	21.90	−20	68
2002	24.26	11	69
2003	24.16	0	67

Source: DTI/BETP, various years.

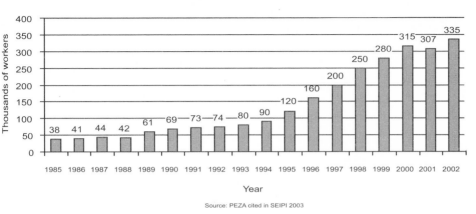

Figure 2.5 Number of workers employed in Philippine electronics

one of the few industry "winners" for the Philippines in the globalizing 1990s. However, behind these few aggregate statistics, there are some important caveats and a more complicated picture of a booming, yet precariously concentrated industry.

First, the industry has had little structural change over the years. The industry is still dominated by wholly owned MNC affiliates and largely concentrated in semiconductor assembly and testing, the low-value-added portion of the global production networks. Despite its rising capital intensity, electronics manufacturing remains highly import and technology dependent with only shallow roots in the local economy. Local Filipino companies mainly do subcontracting jobs from abroad or from locally based MNC affiliates. Semiconductors accounted for an average of 78 percent of electronics exports from 1991 to 2002. While semiconductors continue to dominate, the crash in semiconductor prices in 1997 and cyclical volatility after 2000 have led to the recent decline in its market share.

A second major subsector is computer hardware and peripherals—including hard disk drives and notebook computers—which has been increasing its share of electronics exports, from less than 10 percent in 1991 to 24 percent in 2002. Computer hardware has become the second leading Philippine export behind semiconductors, contributing 11 percent of total industrial manufacturing exports.

Finally, a now tiny portion of the industry includes telecommunications equipment—radios, cellular phones, and telephones—that fell from 9 percent of the market in 1991 to 0.8 percent in 2002.

At an even more disaggregated level, we see that in 2001, 67 percent of all semiconductor exports were devices manufactured from materials imported on a consignment basis (DTI/BTP 2002). This means that more than *32 per-*

Table 2.2. Structure of Philippine electronics exports, 1991–2002 (percent)

Year	Computer hardware	Semiconductors	Telecommunications	Other electronics products
1991	9.8	79.8	9.2	1.2
1992	7.5	78.9	12.0	1.6
1993	6.4	80.0	11.6	2.0
1994	5.1	81.5	10.6	2.9
1995	6.0	83.6	7.9	2.5
1996	8.2	82.2	7.6	2.0
1997	14.1	78.2	5.7	2.0
1998	13.3	78.9	3.6	1.7
1999	15.4	78.9	1.8	1.0
2000	18.4	74.0	.6	7.0
2001	23.1	68.1	.9	7.9
2002	24.3	69.6	.8	5.3
Average	12.6	77.8	6.2	3.3

Source: DTI/BETP various years.

cent of all Philippine exports are entirely dependent on consigned and imported materials, which are assembled, then directly re-exported. These figures clearly show that the Philippine exports are not only dangerously concentrated in electronics, but also that within electronics, the exports are highly dependent on only a few semiconductor products that themselves have virtually no backward or forward linkages within the Philippines.

Companies: Dominance of MNCs

There are a total of eight hundred electronics companies registered with either the Board of Investments (BOI) or the Philippine Economic Zone Authority (PEZA). Numerically, we see that the majority of the firms are Japanese, with Filipino, Korean, American, European, and Taiwanese companies also well represented (see table 2.3). But in terms of value of output, local MNC affiliates dominate the industry, accounting for more than 80 percent of exports. Of the top fifty exporters from the Philippines, forty-two are local affiliates of MNCs. These companies are concentrated around the economic zones in the provinces of Laguna and Cavite just south of the capital (400 firms), in Metro Manila (328 firms), in the special economic zones in Mactan, Cebu (50 firms); in northern and central Luzon at the Subic Freeport and Clark Special Economic Zone (12 firms); and around Baguio City (12).

There are two basic types of manufacturers operating in the Philippines: contract manufacturers and in-house manufacturers. Contract manufacturers

Table 2.3. Electronics firms in the Philippines by nationality

Country of origin	Number of firms	Percent of total
Japan	240	30
Philippines	224	28
Korea	80	10
USA	72	9
Europe	56	7
Taiwan	32	4
Singapore	16	2
Malaysia	16	2
Others	64	8
Total	800	

Source: SEIPI 2003

assemble and test custom-designed circuits and ship directly to their customers or end-users, such as television manufacturers or cellular phone producers. These companies, which may be MNCs or local firms, work mainly with imported materials on a consigned basis. Amkor/Anam is a prime example of a leading MNC contract manufacturer. Most of the 224 Filipino-owned firms are contract or subcontract manufacturers. In-house manufacturers, on the other hand, are wholly owned subsidiaries of MNCs that concentrate on the assembly portion of their mother companies' products. Intel, Motorola, Texas Instruments, Philips, Hitachi, Toshiba, and Fujitsu are all leading firms that have semiconductor assembly and testing facilities in the Philippines.

Imports

While electronics companies have been the leading exporters since the 1990s, they have also been the leading importers. Table 2.4 shows that from

Table 2.4. Share of electronics to total Philippine imports, 1991–1999 (percent)

Subsector	1991	1992	1993	1994	1995	1996	1997	1998	1999*
Computer hardware	1.7	2.3	1.3	1.3	1.6	2.3	3.6	4.4	3.8
Semiconductors	20.9	12.6	13.2	16.0	17.2	18.7	22.1	25.0	25.3
Telecommunications	3.0	2.7	3.0	3.6	4.2	5.5	5.0	2.89	2.0
Other electronics	3.4	3.1	3.4	3.6	4.0	4.1	4.1		
Total	29.0	20.7	20.9	24.4	27.0	30.6	34.9	36.7	34.9

*based on January to July figures
Source: DTI 2000

1992 to 1999, electronics' share of total Philippine imports rose from 20 percent to almost 35 percent. Semiconductors dominate imports, accounting for more than 43 percent of Philippine merchandise imports. However, semiconductors have been a net foreign exchange earner since 1992, while computer hardware has been a net exporter since 1996 (DTI; Austria 1999).

Value Added

Import dependence and weak domestic technological capabilities are the major contributors to the low-value-added level in the industry. In 1992, a study by the electronics industry association showed that value added in Philippine semiconductors was just 11.9 percent. During the 1990s, this has risen, but only slightly. According to a 1997 World Bank study, the average local content is still only about 20 percent in semiconductors. This ranges from 25 percent in simple processes such as printed circuit boards to 15 percent in more complex products like microprocessors. Thus the more advanced processes, which are the bulk of the industry and controlled mainly by MNCs, have the lowest value-added. Given the industry's import and technology dependence and lack of local suppliers, value-added is made up almost entirely of labor inputs.

This is also clear when viewing the industry's contribution to Philippine value-added manufacturing. For all electronics, value-added increased from 6 percent of total manufacturing in the early 1990s to 11 percent in 2000. Thus, although by far the largest exporter, electronics is actually behind food manufacturing, wearing apparel, chemicals, and petroleum refineries in terms of gross value-added (DOLE/BLES 1996; NSCB 2002). Local content has also not grown significantly in more than twenty years, which suggests that small- and medium-sized local suppliers have failed to develop despite the demands of and proximity to large MNCs. The "missing middle," or lack of local suppliers, greatly diminishes the spillover effects of MNCs operating in the country and also lowers the attractiveness of the country to further investments.

But medium-sized local suppliers who were able to raise their skills, technologies, and thus their local value-added through their relations with MNCs were precisely the drivers that helped Taiwan, South Korea, and Malaysia build dynamic and globally competitive domestic electronics industries. By targeting the development of these local suppliers, these countries were able to diversify their industries, acquire advanced technologies, raise the level of value-added, and thus leverage electronics exports for local development. In Malaysia, average value-added has reached about 45 percent while in Taiwan it has reached 75 percent. Local suppliers in these countries were also able to move from simple assembly into more advanced, stable, and profitable types of manufacturing, including original equipment manufacturing (OEM) and even to original design manufacturing (ODM).

Industry Concentration

The overall structure of the industry reveals that among leading electronic exporting countries, the Philippines is the only country where the electronics industry is so highly concentrated on just one sector. Most other countries, such as Malaysia, Thailand, Singapore, China, and South Korea, all have two or three sectors in which electronics manufacturing is concentrated. For example, Malaysia and South Korea, while both led by semiconductors, also have strong computer hardware and telecommunication industries and have also shifted away from labor-intensive assembly and testing. In addition, both countries' semiconductor industries are dominated by domestic companies rather than MNCs (Austria 1999).

The high concentration in semiconductors is not in itself dangerous, as long as world demand for semiconductors remains high and increasing amounts of technology are passed on to local producers and suppliers. But as we noted above, technology transfer to local companies has been slow. The semiconductor industry is also notoriously volatile. The severe contractions of the global semiconductor market in 1996 and again in 2001 caught the industry completely off guard. Because of its high dependence on semiconductors for overall export earnings, the Philippines is particularly vulnerable to such major market collapses.

Philippine State and Investment Policies

The Philippine electronics industry's import dependence, failure to upgrade, and its shallow roots in the local economy can be traced, in large part, to the inability of the Philippine state to guide industrial deepening. Indeed, many have blamed the weakness of the national state for the Philippines dubious status, particularly in the early 1990s, as a "'non-miracle' in a region that was . . . teeming with miracles" (Lim 1999, 15; Hutchcroft 1998). It is no accident that the developmental states of the original "East Asian Miracle" economies of Japan, South Korea, and Taiwan made the electronics sector central to their industrialization strategies (World Bank 1993; Wade 1990; Amsden 1989; Evans 1995; Weiss 1998). And despite continued squabbles over the "success" of the East Asian model and state intervention since the 1997 Asian financial crisis and the spread of greater market liberalization globally, commentators across the ideological spectrum generally agree that states remain a central force in shaping national industrial development (O'Riain 2004; Wade and Veneroso 1998; World Bank 1997; Jessop 2002). Thus in assessing the lackluster performance of Philippine electronics and the role and character of the Philippine state, it is crucial to point out the fundamental differences between the Northeast Asian developmental states and the experiences in Southeast Asia, particularly in terms of state autonomy, institutional capacity, industrial policy, and political context.

Electronics and the East Asian Developmental State

One of the key differences is the reliance on foreign capital and technology. As Jomo (2000, 27) notes, "Whereas [Northeast Asia] successfully employed industrial policy to develop indigenous firms which became internationally competitive, [Southeast Asia] has been primarily reliant on foreign direct investment" (see also Jomo 1997). At the core of the Northeast Asian model were a set of "essentially mercantilist" trade and industrial policies conceived of and carried out by a developmental state and a bank-based financial system bent on constructing national industries in key intermediate and export sectors (Beeson and Robison 2000, 11). Evans (1995), focusing on electronics in his influential comparative study, defines the developmental state as a strong, competent bureaucracy intervening strategically in the economy to construct and improve the country's comparative advantage. He argues that the key to the developmental state is "embedded autonomy," which "combines Weberian bureaucratic insulation with an intense connection to the surrounding social structure," in particular, business elites (Evans 1995, 51). The definition is echoed by others, who focus more specifically on the Taiwanese and South Korean state's "governed interdependence" and "transformative capacity" for guiding industrial change, which requires "coordinated investment, diffusion of innovation, and generally ensuring permanent upgrading of the industrial portfolio, especially in the tradables sector" (Weiss and Hobson 2000, 62; Weiss 1998). Finally, O'Riain (2004), drawing on his study of the Irish state and its role in promoting an indigenous software industry, develops a more decentralized, nonbureaucratic "post-Fordist" model of state intervention, which he calls the developmental network state (DNS). While sharing the overall goal of the more bureaucratic model—the development of regional systems of innovation and national economic growth—the DNS focuses on the development of local public and private sectoral capacities and relational assets to "better" connect a region to the global economy through the "centralized coordination for the overall integration of its decentralized networks" (O'Riain 2004, 33).

As noted above, such "permanent" upgrading and regional "systems of innovation" are particularly important in the advanced electronics sector, where technological leadership is the key to competitiveness. Vital for the South Korean and Taiwanese cases was continued state leadership that ensured that local firms gained a foothold in the international industry and which then lead this upgrading. For example, the South Korean state directly targeted electronics as one of its "five pillars of wizardy" or five key industries in its industrialization plan in the 1960s and 1970s that began with subcontracting from American and Japanese firms at the strategically designed Kumi Electronics Complex (Woo 1991, 133). Similarly, Taiwan targeted electronics very early on, creating one of the first export processing zones in Kaoshiung in 1966 to draw multinational firms that controlled much of the technology (Wade 1990; Weiss 1998). However, South Korea was equally determined to move quickly out of subcontracting though a concerted, state-led effort. Specifically, the Korean state, through preferential bank-lending policies, focused on direct support of the enormous Korean conglomerates, or *chaebols,* the provision of sophisticated infrastructure, technological targeting in the computer and semiconductor industries, increased sector-specific research and development, heavy investment into training, and not insignificantly—the suppression of organized Korean labor (Woo 1991; Ernst and O'Connor 1992; Koo 2001). The result: South Korea, which had no semiconductor manufacturing industry until the early 1980s, has become the world's leading producer of sophisticated dynamic random access memory (DRAM) semiconductor chips (Austria 1999). Likewise, Taiwan invested massively in research and technology institutes, bid out technology transfer contracts, and promoted small- and medium-sized enterprises (SMEs), both state- and privately owned, to transform Taiwanese firms from assembly subcontractors into the world's dominant manufacturers of computer motherboards and capital-intensive silicon wafer fabrication (Austria 1999; Linden et al. 2001).

The Philippines: From "Strong" to Regulatory State

The Philippine state does share some similarities with Taiwan and South Korea and in many ways, it tried to emulate the industrialization strategies of its northern neighbors. The Philippine state has certainly had periods when it was strong, interventionist, and authoritarian and also tried to stimulate investment through the building of export processing zones. But as Peter Evans points out, "The appropriate question is not 'how much' [state intervention] but 'what kind'" (Evans 1995, 10). Indeed, despite high expectations since the 1950s, a highly educated populace, abundant natural resources, and at least a vision of export-led industrialization, the Philippines found itself by the 1970s in a "developmental bog" and dubbed the "sick man of Asia." Nevertheless, just as the meteoric rise of the South Korean and Taiwanese economies are better understood in a broader historical, political-economic, and global security context, so too are the "failures" of the Philippine state and its inability to promote genuine national development (Woo 1991; Wade 1990).

Hutchcroft (1998, 52) characterizes the Philippines as a "patrimonial oligarchic state," defined as a polity in which "the dominant social force has an economic base largely independent of the state apparatus, but the state nonetheless plays a central role in the process of wealth accumulation. . . .

In contrast to bureaucratic capitalism, where the major beneficiaries of rent extraction are based within the administrative apparatus, the principal direction of rent extraction is reversed: a powerful oligarchic business class extracts privilege from a largely incoherent bureaucracy." The roots of the contemporary "weak" Philippine state and its capture by groups of powerful oligarchs can be traced to its colonial history and its fitful integration into the capitalist global economy since the "long 19th century" (McCoy and de Jesus 1982, 6). During this period, the commercialization of Philippine agricultural goods such as sugar, rice, hemp, and tobacco helped create a powerful, regionally based landlord class that remained independent of the central Spanish colonial state. By 1898, when the United States wrested colonial control, these landlords, or *caciques,* had become so powerful that the American colonizers struck a political bargain with them in order to win local support for the new colonial regime. Buoyed by the support of the colonizers, these agro-exporters used their positions in the newly created national assembly to strengthen their political power, using their gatekeeping control over a weak central state as a key to accumulating rents, diversifying their businesses, and shutting out rivals (Hutchcroft 1998). This group, composed of a relatively small number of key elite families, astutely read the geostrategic importance of the Philippines in the Asia Pacific, and the continued dependence of the United States on their support for legitimacy, which was needed to maintain U.S. military bases. Thus, by supporting the strategic aims of the U.S. government, they were given free rein to block any meaningful land reform or consistent economic development policies and cement a system of "patrimonial plunder" that has led to a historically anemic state bureaucracy throughout most of the twentieth century (De Dios and Hutchcroft 2003; McCoy 1993).

The power of these elite families acting in defense of their business interests greatly influenced the industrialization policies of the state. In the early 1950s, the newly independent country was reeling from a balance of payments crisis, which led to protectionist measures aimed at both resolving the crisis and promoting industrial development. But the import substitution industrialization (ISI) plan proved to be yet another state project that was quickly captured by the oligarchs, who used their access to the state and high tariff barriers to gain control over whatever import-dependent light manufacturing they could, with little regard for a systematic industrialization strategy (Rivera 1994; Kang 2002). By the late 1960s, the incoherence of the development policies and the dominance of the now-diversified family conglomerates led to the stagnation of the ISI experiment and another balance of payments crisis.

When Ferdinand Marcos became president in 1966, he promised to dismantle the oligarchs' stranglehold and usher in a new era of state-led, export-oriented growth. To this end, he neutralized some of the leading oligarchs—such as the Lopez and Osmena families from the southern islands—and surrounded himself with American-educated technocrats who looked toward the rising economies of South Korea and Taiwan for inspiration. However, after centralizing power, Marcos intensified the system of "patrimonial plunder," enriching himself and a set of new oligarchs or cronies rather than implementing a sound development strategy (Hutchcroft 1998). Facing yet another economic crisis as well as a constitutional term limit law, Marcos declared martial law in 1972, dissolving the congress and suspending the constitution, justifying his actions as necessary for implementing the new export-led plan of national development (Hawes 1987).

Much of the pressure to come up with an export-led development plan came from the World Bank, the International Monetary Fund (IMF), and a bevy of creditors on whom Marcos was becoming increasingly dependent to solve the Philippines balance of payment woes (Bello, Kinley, and Elinson 1982). But while inspired by the successes in South Korea and Taiwan, the World Bank and IMF clearly held their own neoliberal vision of the "economic miracle" and "like Narcissus, the World Bank [saw] its own reflection in East Asia's success" (Amsden 1994, 627). Their specific interpretation of East Asia's export-oriented industrialization (EOI) strategies did not follow the interventionist "developmental state" model, but rather, advocated a restructuring of the economy toward more free market policies. They urged the Philippine government to dismantle the protectionist ISI policies and replace them with more open policies that would be friendlier to foreign direct investment (FDI) and could exploit the country's comparative advantage in the low-wage, export sector. To this end, the government ushered in a series of reform packages, including a 60 percent devaluation of the peso, duty-free imports of raw or semifinished materials for subsequent exports, liberalization of FDI and trade laws, and lowering the minimum foreign investment floor.

In addition to investment incentives, another centerpiece to the new plan was the establishment of an export processing zone program modeled after Kaoshiung, Taiwan. First, President Marcos created an additional package for foreign investors available only in the zones, allowing 100 percent foreign ownership, duty- and tax-free production, unrestricted repatriation of profits, and the right to borrow money locally with Philippine government guarantees (Villegas 1989). Then, he brought online the first export zone, the state-owned Bataan EPZ, on the northern edge of Manila Bay, complete with dedicated power, prefabricated factory buildings, and worker housing. Yet just as important as infrastructure and monetary incentives was the creation of the proper political climate for foreign investment. The Bataan EPZ opened just two months after President Marcos declared martial law and vastly expanded the coercive power of the state. Under martial law, Marcos rewrote the national labor code to favor employers; banned all strikes in "vital industries," specifically in the export manufacturing sector; and "nationalized" the trade union structures in order to control militant labor and effectively wipe out more than 5,500 of the 7,000 registered unions and labor federations. The connection between heavy-handed labor control—particularly within the new EPZs—and promotion of foreign investment was obvious. As the then-chairman of the Board of Investments stated candidly, "The logic of foreign investment to participate in the generation of labor-manufactured exports is clear and incontrovertible, but the country needed martial law to attract such investment" (Paterno 1973 cited in Bello, Kinley, and Elinson 1982, 139). And as a confidential report by the World Bank explains, this was not an unrecognized advantage for implementing their suggested economic reforms: "The abolition of Congress provided the government with almost absolute power in the field of economic development" (World Bank 1976, cited in Bello, Kinley, and Elinson 1982, 139). Thus the success of the zones and the liberalization of the export sector were predicated on maintaining strong state control, particularly over increasingly militant labor. It was at this early stage, in 1974, that two of the largest electronics multinationals—Intel and Texas Instruments—set up assembly plants in the country. The Bataan EPZ proved initially successful in drawing foreign investments, and the EPZ program expanded with the opening of new public EPZs in Mactan, Cebu, in 1978; Baguio City in 1979; and Rosario, Cavite, in 1986.

Yet despite these early moves to promote exports and foreign investment, the Marcos government did not make a fundamental shift away from the patrimonial policies of the past. This was in large part because Marcos and other oligarchs, who through protection and monopolistic position could exploit a move into the export sector, were nevertheless in poor shape to compete internationally with vastly more efficient foreign multinationals. Rather, Marcos and his cronies found it easier and more lucrative to establish formal export and foreign-investment-friendly policies and use the rhetoric of its Western-trained technocrats to keep funds flowing from private banks, the World Bank, and the IMF. During the martial laws years, between 1973 and 1981, the World Bank alone injected some $2.6 billion into sixty-one major projects (Bello, Kinley, and Elinson 1982). The easy access to external funds and the Philippines' continued geostrategic position for the Americans allowed Marcos to avoid making the tough policy changes required for a comprehensive, export-led industrialization push. Instead, with the declaration of martial law and the concentration of state power, Marcos pursued a contradictory but politically savvy, two-pronged strategy: he created a limited export platform or enclave economy to please foreign investors and development agencies while maintaining a highly protectionist domestic economy and coercive state to fend off political rivals and enrich himself and his closest allies (Kang 2002).

In some ways, the Philippines did mimic the early model of the Taiwanese and Koreans: using the broad powers of the state, it developed an enclave economy centered on foreign investment in free-trade export zones. However, unlike the South Koreans and Taiwanese, the Philippines did not connect these enclaves to a broader, more comprehensive industrialization plan to promote local industry and spur competitive, indigenous firms. The persistent clientelist character of the government meant that, because of political manipulations, intervention in the economy was often inefficient or unproductive. Compared to the East Asian cases of South Korea or Taiwan, the Philippines remained a far cry from a strong state that could *competently* intervene in markets and guide economic development without industrial policies being captured by rent seekers.

After the severe economic and political crisis that finally toppled Marcos in 1986, the country began moving even further from the strong, development state model, both because Filipinos were understandably wary of a "strong" central state and—more significantly—because the state had become increasingly dependent on the World Bank, IMF, and foreign creditors for further bailouts. The ouster of President Marcos through the "People Power Revolution" rid the country of a brutal dictator but left the country saddled with an enormous external debt of more than $26 billion (Pineda-Ofreneo 1991). By the early 1980s, the World Bank and the IMF had already grown weary of Marcos's promises and had begun making further loans contingent on actual reforms. The Philippines was one of the first four countries to receive the World Bank/IMF's new structural adjustment loans (SAL) in 1980, which for the first time tied continued credit to a liberalization program of macroeconomic changes set by the creditors.

Pressured to restructure the economy under the IMF's program loans and desperate for both more credit and the return of foreign investment, President Corazon Aquino chose a "model debtor" strategy, sacrificing domestic investment in growth and poverty alleviation to servicing the external debt and further liberalization of the economy (Lindsey 1992 cited in Bello 2000). Between 1986 and 1989, there was a negative outflow of financial resources, with total debt payments reaching more than $3 billion a year and external debt servicing alone siphoning off 50 percent of the entire government budget (Bello 2000). Meanwhile, the state continued to liberalize the economy, further devaluing the currency, tightening fiscal and monetary policies, dismantling quantitative restrictions on imported consumer goods, slashing tariffs, and easing restrictions on foreign investments. In 1989, the Philippines signed a Letter of Intent with the IMF, stipulating increased taxes, state assumption of private debt, removal of government rice subsidies, and reigning in of other government expenditures (Pineda-Ofreneo 1991). To draw back foreign investors, the government also expanded and streamlined the foreign incentive packages through the 1987 Omnibus Investments Code.

Added incentives included an income tax holiday, tax- and duty-free importation of capital equipment, simplified customs procedures, and access to bonded warehouses. While such moves did succeed in dismantling many of the protectionist policies of the past, the "shock therapy" led to economic stagnation, increased inequality with nearly half of all Filipino families living below the poverty line, and a state in a weak position to pursue any economic development program not dependent on foreign investment.

The Regulatory State

In 1992, Fidel Ramos was elected president in part on his promise to transform the Philippines into the next Asian Tiger. However, his vision for "Philippines 2000" was refracted through a particular ideological lens and a firm belief that a globally competitive Philippines had to make itself more attractive to foreign investment through freer trade and investment, deregulation, and extensive privatization (Rocamorra 1994). Ramos and his economic team broke up government or crony-dominated monopolies in telecommunications, energy, and agriculture, further reduced tariffs and import restrictions, restructured the central bank, and eliminated most capital controls. Thus his administration began a systematic neoliberalization of the economy and the transformation of the Philippine government into what is best characterized as a regulatory state (Jayasuriya 2003; 2000). The regulatory state in the Philippines is an extension of Southeast Asian state, which as noted above, has always relied more on foreign direct investment than on coordinated industrial policies. Indeed, the Philippine state under Ramos has gone in the opposite direction of the Northeast Asian developmental states. As noted above, South Korea and Taiwan had used a mercantilist policy of restricting imports while at the same time developing the domestic capacity to export. But the neoliberal Philippine state has embarked on what de Dios and Hutchroft (2003, 73) have called a "reverse mercantilist" policy: allowing the unfettered import of both final and intermediate goods but developing no coherent industrial deepening strategy to promoted higher value-added exports. This has led to a persistent dependence on both imports and foreign direct investment to fuel the domestic economy but without a strategy to upgrade the Philippines' position in the global division of labor. Thus, unlike elements within the "post-Fordist" Irish state, which used a crisis in the FDI-driven development model to push for institutional reforms toward increasing spending on education, research, and development, and the fostering of an indigenous system of innovation, the Philippine state actors actually intensified their pursuit of FDI by offering a new kind of "neoliberal localization" (O'Riain 2004; Brenner and Theodore 2002).

What defines the regulatory state, then, is not the wholesale decline of state power under the pressures of globalization, but rather its shifting emphasis on promoting interlocal competitiveness, safeguarding the market, and providing nonmarket forms of coordination and place-specific assets to support more sophisticated forms of capital accumulation (Brenner and Theodore 2002). Since the early 1980s, with the increasing influence of the World Bank and IMF on both the economy and the state, the Philippines had already begun experiencing what Peck and Tickell (2002, 39) have called the "asymmetric scale politics of neoliberalism" in which, "local institutions and actors were being given responsibility without power, while international institutions and actors were gaining power without responsibility."

Thus neoliberalization is not simply the achievement of free trade and the retreat of the state, but a process in which an active state intervenes at a variety of levels to secure and insulate markets and their regulatory institutions from other political influences (Jayasuriya 2000). This dynamic process involves the simultaneous "roll back" of certain state interventions—such as state subsidies, state-owned enterprises, and national protectionist policies—and the systematic "roll out" of new neoliberal institutions and policies aimed at "politically constructing" certain markets or parts of the economy, often on a local or regional, not national level (Peck and Tickell 2002). Some intellectual leaders within the Philippines have clearly articulated this notion. Expanding on the call for "good governance" and not simply free markets, the president of the Development Academy of the Philippines noted that "despite calls for the redirection of power towards markets and away from governments, the state has a central role to play—not just its many traditional and contemporary undertakings but *through its leverage over other key players within society*" (cited in Gonzalez 1999, 236 emphasis added).

Instrument of the Regulatory State: the Philippine Economic Zone Authority

The shift to a regulatory state is probably best exemplified by the establishment of the Philippine Economic Zone Authority (PEZA) in 1995. As noted above, FDI in the electronics sector had long been promoted as integral to the government's development strategy, and the state relied mainly on extending its long list of incentives package to lure investors. But through the Special Economic Zone Act, the state aggressively reorganized its key coordinating institution, creating PEZA and paving the way for restructuring, expanding, and privatizing the zone program and all that it provided. First, the new law further expanded incentives, prompting no less than the World Bank to state that "the Philippines' ecozone incentives are easily the most generous and flexible set of incentives available anywhere" (World Bank 1997, 88).

Specifically, the PEZA incentives include:

- An Income Tax Holiday or exemption from Corporate Income Tax for eight years, after which enterprises have the option to pay a special 5 percent tax on gross income in lieu of all national and local taxes.

- Exemptions for duties and taxes on imported capital equipment, spare parts, supplies, and raw materials.
- Domestic sales allowance of 30 percent of total sales.
- Exemption from wharfage dues and export taxes, impost, and fees.
- Additional 50 percent tax deduction of the total costs of training for labor and management.
- Access to bonded manufacturing warehouses.
- Simplified customs procedures.
- Net operating loss carryover and accelerated depreciation.

But beyond incentives, PEZA also moved away from large, publicly owned zones, with their poor coordination and antiquated infrastructure, to a new model of smaller, privately owned zones. The new special economic zones reflected the neoliberalization strategy of "glocal" developmentalism or, "the promotion of global economic competitiveness within strategic subnational territorial sites such as urban regions and major industrial districts" (Brenner 2000, 370). In many ways, these new zones were simply an intensification of the "enclave economy" approach, which began in the 1970s. But to match the more complex competitive demands of advanced manufacturing investors, these zones have expanded their appeal: they now target specific sectors—notably electronics—and provide state-of-the-art infrastructure such as dedicated and uninterrupted power and water supplies, waste water treatment facilities, high-speed and reliable communications, and easy access to air transportation networks. PEZA also created regional and on-site offices that served as the lead state institution, or "One Stop Shops" for investors that helped insulate investors from demands of competing state offices, such as the Department of Labor and Employment (DOLE). At the local level, PEZA became intimately involved in everything from how the zone was constructed and laid out, to labor recruiting, to the dispersal of working housing, to the management of labor relations, and labor control in the community. These activities at the local level—part and parcel of a coordinated policy of "strategic localization"—will be presented in much more detail in chapter 4.

The incentive package, streamlined customs procedures, the provision of solid infrastructure, and the coordinated community involvement down to the local level all helped Philippine zones take off. By the mid 1990s, the Philippines was second only to Taiwan in the region in terms of the absolute levels of total exports, direct employment, investment, and number of enterprises. The Philippines also became one of the top-five export zone programs in the world (World Bank 1997). Electronics multinationals have been the major players in the success of Philippine economic zones. From 1995 to 2003, firms engaged in electronic components and products and electrical machinery accounted for 68 percent of all investors and about 70 percent of total zone investments. In 2003 there were 294 electronics firms operating in the PEZA zones. Total direct and indirect employment in the zones reached over 900,000 in 2003, while manufactured output from the zones rose from $2.7 billion in 1994 to over $27 billion by 2003 (PEZA 2004).

However, despite the "rollout" policies of the regulatory state, other developmental policies to promote industrial deepening were notably absent and many feel that the incentive package have been far too generous. The tax exemptions and deductions offered by PEZA are only matched in the region by Thailand while the duty-free importation of nearly all project-related materials, including production inputs and without time limit, is unrivaled. Other countries limit duty-free importation either to capital equipment, production-related items, and items not available locally, or provide this incentive only on initial investment. The World Bank even noted, "Duty-free imports of all project-related inputs is excessive, going well beyond international norms, and probably unnecessary for the competitiveness of most enterprises" (World Bank 1997, 89).

While the incentives appear to have successfully lured leading multinational electronics manufacturers to the Philippines, the incentive package lacks clear policies that would help boost local value-added, such as backward and forward linkage programs to identify and develop local component and raw material suppliers (McKendrick et al. 2000).

Government investment and regulation policies are at least in part responsible for industry concentration and its shallow roots in the local economy. In countries where the early entrance in electronics assembly was successfully leveraged to new levels of sophistication and industrial deepening, the states all played fundamental roles. In general, Taiwan, South Korea, Singapore, and Malaysia took forceful policy steps toward "paving the high road" with more dynamic and locally integrated electronics production while at the same time "closing off the low road" option of simply sweating cheap local labor. These countries all invested large amounts of public money and political will into developing the local capacity of firms and manpower to absorb MNC-provided technologies, develop original equipment and design manufacturing capabilities, and move upward in the value chain. At the same time, these countries also "disciplined" MNCs once established, by eventually lowering tax incentives, holding them accountable for certain performance standards, requiring investments into government-regulated training programs, and requiring a minimum level of local sourcing (Ernst 2002).

The Philippines, fearful of scaring off investment, has simply maintained or even expanded its "all carrots and no sticks" strategy that combines an overly generous incentive package with enclave infrastructure and selective labor regulation. While the government has been successful in attracting some of the largest electronics MNCs in the world, it has not been vigilant in promoting and regulating technology spillovers to take advantage of their presence.

The Philippine Labor Force and Labor Market Regulations

As noted above, electronics multinationals put labor as one of the top criteria for investment. But as the nature of market competition has changed, high-tech producers are no longer simply looking for cheap "nimble fingers." Instead, leading firms with large capital investments in the latest production technologies are increasingly searching for a labor force that is not only cheap, but also productive, skilled, and "flexible." As the World Bank and many multinationals have recognized, the Philippines' main competitive advantage is its range and abundance of skilled, English-speaking, and relatively inexpensive managerial, technical, *and* production labor (World Bank 1999). These characteristics have been crucial for drawing advanced investments.

But as competition has become more demanding, firms are also not content in simply responding to labor market signals nor do they treat the labor market as exogenous. Thus it is important to take a much broader view of the labor market, which is best defined by Peck (1996, 160) as "the social relations surrounding the sale and purchase of labor power." As we will see below and in subsequent chapters, the importance of labor and labor control to both competitiveness and locational decisions has meant an increasing intervention by both firms and the state in labor market regulation.

Philippine Labor Market

Two of the primary advantages Filipino workers (especially women) tend to have over workers in other developing countries are their high level of educational attainment and ability to speak English. Due to the legacy of American colonialism, the Philippines is now the third largest English-speaking country in the world, with a quite high basic literacy rate at 92.6 percent and a female literacy rate of almost 95 percent. This compares favorably to China, Malaysia, and Indonesia, with literacy rates of about 90 percent; and India, at only 65 percent. It also boasts secondary level school enrollment of 79 percent (compared with Malaysia at 58 percent), and a college level enrollment of 27 percent (versus 10 percent in Malaysia) (NSCB 2004; World Bank 1999). In fact, after Singapore, the Philippines has the highest enrollment at the secondary and tertiary levels in Southeast Asia, the highest percentage of engineering students, and is by far the largest producer of engineers: more than 200,000 enrolled, or four times the number of Thailand (McKendrick et al. 2000). As the Philippine electronics industry association likes to boast, the Philippines produces 100,000 engineers, IT, and technical graduates each year (SEIPI 2001).

Despite its impressive education numbers, the Philippines remains a poor country with a large labor surplus. The average annual family income is only

Table 2.5. The Philippine labor market, key indicators

Total population	84.2 million
Annual population growth rate	2.36 percent
Labor force	36.5 million
Unemployment	13.7 percent
Underemployment	18.5 percent
Poverty rate	34 percent

Source: NSCB 2004

thirty-six hundred dollars and more than 34 percent of all Filipino families fall below the poverty line. The population, now at 84 million, is also growing at a rate of 2.3 percent per year, making it very difficult for the country to absorb the 740,000 new entrants and reentrants into the labor market each year. The unemployment rate, at nearly 14 percent, is high, while the numbers of underemployed—those that have some work but would like more—is still higher, at 18.5 percent. It is also important to point out that the unemployment rate generally rises with the level of education.

In terms of sectoral employment, nearly half of all workers, 49 percent, are employed in services, while 35 percent are in agriculture, and just 16 percent are in industry. In fact, despite increasing industrial investments, the manufacturing sector's share of total employment has actually *declined* in the last thirty years, from 12 percent in 1970 to less than 10 percent in 2004—the lowest level among East Asian countries (NSCB 2004; Herrin and Pernia 2003).

In Region IV, southern Tagalog, just south of Manila, where 60 percent of the major electronics firms are located, wage rates remain low and poverty widespread. For example, in 1998, despite the boom in private and special

Table 2.6. 2003 Unemployment rate (%) by highest grade completed

Total	11.4%
No grade completed	2.4
Elementary	11.3
Some high school	14.4
High school graduate	27.1
Some college	17.8
College graduate	16.7

Source: NSO 2004

economic zones in the region, unemployment was above the national average at 10.1 percent, underemployment stood at 18.3 percent, and more than
25 percent of the population lived below the official poverty line. The nom-
inal and official minimum daily wage rate in Region IV in 1998 was P188
or less than five U.S. dollars per day. Yet agricultural wages were only P157
a day, and rural wages were as low as P118 per day. While electronics firms
paid the prevailing minimum wage rates or higher, they were not substantially higher than average national or regional wages (BLES 1999).

The state of the overall labor market is important, particularly when assessing the Philippines' chances of emulating its Northeast and successful
ASEAN neighbors. Tight labor markets in South Korea, Taiwan, Singapore,
and Malaysia provided upward pressure on wages, which helped convince
both local governments and MNCs that production would need to be upgraded to higher value-added niches to stay globally competitive. While the
electronics industry in the Philippines has seen rising employment, it still represents only 10 percent of manufacturing employment and less than 1 percent of total employment. Thus there is little likelihood that the recent boom
in electronics will lead to upward pressure on wages as it did in Malaysia.
And slack labor markets make it more tempting (and viable) for firms to continue to focus on cost-cutting strategies of flexibility. Surplus labor may also
make it possible for firms to embark on upgrading without having to directly
share the benefits with workers. A broad study of industrial clusters in developing countries showed that in a context of surplus labor, larger firms
were able to successfully upgrade their technology and production while still
maintaining low wages (Nadvi and Schmitz 1994).

Another key factor influencing foreign direct investment is the level of
worker militancy and unionization rates (McMillan, Pandolfi, and Salinger
1999). The Philippines has a history of militant and political unionism, but
the labor movement since the early 1990s has been in relative decline. The
rise in militant unionism grew in response to heavy labor repression under
Martial Law and as part of the growing movement against the Marcos
regime (Scipes 1996). The labor upsurge culminated in the founding of the
openly militant Kilusang Mayo Uno (KMU) labor alliance on May 1, 1980
(West 1997). By the mid 1980s, 121 labor federations claimed a total membership of over 1.6 million workers. These numbers have only grown: in
1990, unions claimed a membership of over 2.2 million, and in 2001 the total claimed membership by the 166 labor federations and 9,169 registered
company-level unions stood at 3.7 million workers, or roughly 11 percent
of the workforce and 26 percent of the formal sector wage and salary earners. Yet despite the growth in claimed membership, workers actually covered
by collective bargaining agreements (CBAs) has been much lower and in decline since the early 1990s. From 270,000 covered in 1983, the number of
workers covered by union contracts peaked in 1993 at 608,000, but has

since declined. By 2003 the numbers were far from impressive: only about 550,000 workers, or 3.6 percent of wage and salary earners, and a paltry 1.5 percent of the labor force worked under a union contract (BLES various years). In terms of militancy, strikes escalated during the economic and political turbulence of the 1980s, going from 62 in 1980 to 282 in 1984 to 371 in 1985, and peaking in the crisis year of 1986 at 581. By that time, Manila had earned the moniker, "strike capital of Asia" (ILS 1997). But actual strikes or lockouts have also decreased steadily, dropping to 183 in 1990, 97 in 1995, and by 2003 strikes had fallen well below even the 1980 level, at only 38 (BLES various years).

The Electronics Industry Labor Force

Since the late 1980s, the return of political stability and the decline in union militancy in the Philippines, coupled with rising wage rates in Northeast Asia, Malaysia, and Singapore, led more MNCs to invest in the Philippines. The lowest end assembly work has been shifting to the lowest wage countries, namely China, Indonesia, and Vietnam. However, while wages are indeed much lower than in the Philippines, these countries—so far—lack a workforce with the high literacy levels and English-speaking abilities that are important in the more sophisticated assembly and testing manufacturing, such as in microprocessors.

English-language ability is particularly important in semiconductors as ramp-up time of new facilities depends on quickly training the labor force and most training materials are in English. MNCs in the country claim that training times in the Philippines are among the fastest in the region, taking only six to eight weeks.

Wage Rates

Of course, labor costs are a crucial factor, but advanced manufacturers must now look across an entire range of unskilled, technical, and managerial labor. It is in providing this range of labor that the Philippines proves highly competitive. First, production operators are abundant and wage rates for operators are also relatively low: about $150 to $200 a month, compared with $300–350 in Malaysia, the Philippines' main direct competitor. Wage rates for production operators are still crucial in cost accounting, since operators make up 82 percent of the electronics industry's workforce. Turnover rates for production operators are also low, about 8 percent a year, due in part to the surplus labor market (BLR/DOLE 1999).

For technical positions, engineers are not only relatively cheap but also quite skilled and abundant. Again, tight labor markets for engineers in Malaysia have led to companies having to use expensive expatriate engineers, often Filipinos. New engineers earn about $400–500 per month in the Philippines, while in Malaysia the cost is much higher: $800–1,000 per

Table 2.7. Comparative Asian labor ratings

Country	Skilled			Unskilled			Average
	Quality	Cost	Supply	Quality	Cost	Supply	
Philippines	**2.54**	**2.36**	**0.86**	**2.62**	**2.62**	**2.92**	**2.32**
Vietnam	1.67	0.50	1.00	5.00	2.50	8.00	3.00
China	5.17	1.68	1.90	6.52	3.66	6.88	4.25
Indonesia	5.25	1.29	1.33	6.85	4.83	7.85	4.53
Thailand	5.51	2.90	3.25	6.21	5.13	7.41	5.07
Malaysia	4.00	4.25	6.00	5.00	5.75	5.71	5.45
Singapore	5.51	6.10	5.15	4.52	6.41	6.85	5.84

Source: Political and Economic Risk Consultancy, Ltd. 1997

month. Production supervisors are also relatively cheap in the Philippines; available at $450–600 per month compared with $1,300 in Malaysia (SEIPI 2001).

Finally, the Philippines boasts some of the highest-quality managers in the world. The 1998 World Competitiveness Yearbook ranked Philippine management as the second most competitive in a worldwide survey of forty-six developed and developing countries. A Filipino production manager costs $1000–1500 per month compared to managers in Malaysia who can command $3600 per month. Overall, Philippine labor ranks as the most competitive in Asia in terms of quality, availability, and cost, well ahead of Vietnam, China, Indonesia, Thailand, Malaysia, and Singapore (see table 2.7).

Characteristics of Work and the Workforce

Beyond the aggregate employment and wage statistics are the particular characteristics of the electronics industry workforce. The following analysis draws primarily on an unpublished survey of 124 electronics firms conducted by the Department of Labor and Employment's Bureau of Labor Relations and covering a broad range of data, including workforce demographics as well as labor-management relations and workers' benefits (BLR/DOLE 1999).

GENDER Women account for 73 percent of the workforce in the Philippine electronics industry. This is consistent with data and studies from the economic zones in the Philippines and other developing countries that show a disproportionate share of women workers in labor-intensive industries. As studies from around the world have shown, management prefers women for these types of production jobs, which is both based on and helps reproduce gender stereotypes associating product assembly with "feminine work."[1] First, women are considered more manually dexterous and careful, particularly regarding small parts assembly. Second, women are regarded as more diligent and patient, which is deemed necessary because of the long hours of handling small parts and repetitive work. Third, women are considered easier to manage, more docile, and less likely than male workers to join trade unions. Finally, women are also assumed to be more willing to work for low wages and have a higher rate of voluntary turnover because of their supposed status as "secondary wage earners" in the family. For these and other reasons persistent among firms and particularly among human resource departments, women continue to dominate the industry and thus continue to feed the pervasive stereotypes that have lead to the branding of electronics assembly as "women's work" (Chant and McIlwaine 1995; Chow and Hsung 2002; Elson 1999; Freeman 2000; Lee 1998; Parrado and Zenteno 2001). How gender is "produced" on the shop floor and in the labor market and how gender politics play out in practice will be taken up in subsequent chapters (Salzinger 2003).

There is also strong gender segmentation within firms, creating horizontal as well as vertical obstacles to mobility. While overall the workforce is 73 percent female, as we will see in the four case studies, production operators in large firms are often 85 to 95 percent women. Yet men dominate other, more prominent positions—for example, technicians and engineers. In interviews, industry HR managers and recruiters attest this segregation to the lack of qualified female technical and engineering graduates available. Whatever the reason for the bias, there is a clear gender association between higher technology and technical skills with male workers, which will be explored further in chapter 3. This association has important implications in terms of industrial and job upgrading and the potential for women operators to improve their positions.

AGE A second important characteristic of the electronics industry is the young age of its workers. Seventy-eight percent of the workforce is between

---

1. The most durable of the early arguments regarding women, work, and export manufacturing has been the "exploitation thesis," which holds that transnational producers, in search of cheap and docile assembly hands, employ young single women to maximize control and profitability (Safa 1981). Challengers to the exploitation thesis have countered with evidence of women's socio-economic empowerment through formal employment (Tzannatos 1999), have pointed to women's agency and resistance in the face of transnational capital (Wolf 1992), and have suggested that women workers may form the nucleus of a working class movement (Lin 1987). However, debating the merits of employing "cheap" women's labor may only help reproduce and normalize women's subordinate status rather than explain it. Salzinger (2003, 16), drawing on a poststructural feminist perspective, argues for a more nuanced understanding of how gendered meanings of "feminity" are constructed in production, suggesting that "docile labor cannot simply be bought; it is produced, or not, in the meaningful practices and rhetorics of shop-floor life."

eighteen and thirty years of age and 17 percent are between thirty-one and forty years of age. One reason behind the very young profile is a preference by companies for "fresh" high school or college graduates. Young workers are assumed to have better health, particularly perfect eyesight, which is essential for many production positions. They are also thought to be more productive. But the most important reason for hiring young workers is their lack of previous work experience.

Unlike the garment industry, which often prefers to recruit skilled workers, electronics companies overwhelmingly prefer workers with no previous work experience. The most important criteria is not pre-existing skills but "trainability." This is in part because skills are often firm-specific and companies prefer to do to their training based on their own needs. Firms also prefer those with no previous experience because they find them easier to train and inculcate with company-specific values and culture. Many recruiters liken recent graduates to "fresh clay": easier to mold and shape than those whose work habits and ideas have already "hardened."

EDUCATION Workers in the industry also have a high level of education. Sixty percent of industry workers are high school graduates while 36 percent have college degrees, and 3 percent have vocational or technical coursework. Generally, the companies are looking for workers with a relatively high level of intelligence who can be quickly and easily trained. They also prefer those with a command of English, since most training materials and technical manuals are in English. Recent graduates or school leavers are also preferred because it is often thought that disciplined behaviors learned and practiced in school—punctuality, good study habits, and obedience to authority—is still "fresh" in the minds and habits of young graduates.

For many years, the basic requirement for production workers was a high school degree. But with the trend toward higher technology and increased capital intensity, many companies now require (or at least prefer) workers with at least two years of college coursework, preferably in a technical field. In some of the most technically advanced production in a few MNCs, production operators with a college degree in engineering are sought.

REGULAR VERSUS NONREGULAR WORKERS The changing technical requirements and nature of production in electronics is also reflected in the high level of regular or permanent versus nonregular terms of employment. While the trend in other manufacturing industries, such as garments, has been a shift toward external flexibility or increased contractualization and casualization of work (Torres 1993), in electronics, the movement has been in the opposite direction (albeit nonregular employment is still widely practiced). According to a detailed survey of manufacturing industries in 1990, 52 percent of electronics firms surveyed employed only regular workers

(Windell and Standing 1992). This figure was the highest among the twelve manufacturing industries studied. A more recent survey showed that 73 percent of workers in the industry were regular, permanent employees, while 16 percent were contractuals or apprentices, and 11 percent were probationary (BLR/DOLE 1999).

The trend toward "regularization" rather than casualization is best explained by the growing need for a more trained, stable, and committed workforce. The increased demand for high quality and fewer defects has meant that many MNCs must train workers for up to three months before they achieve sufficient productivity levels. Given their increased investments (of both time and money) into each worker, firms want to retain them for longer periods and provide workers with a more stable environment to build work loyalty.

Despite the positive general trend, casualization and contractualization still persist in the industry, especially among contract and subcontract manufacturers. A survey by the DOLE in 1997 revealed that 45 percent of electronic firms that reported to have surplus labor employed contractual and casual employees.[2] The reasons given for hiring casual and contractual workers are instructive. Ninety percent of firms cited the need to respond to fluctuating demand and market volatility. This clearly reflects the dependence of the industry on export markets and world prices. The other main reasons demonstrate the industry's general attitudes toward the use of static flexibility measures: 83 percent (the second most cited reason) stated they hire casuals and contractuals to "prevent workers from organizing" labor unions, while 79 percent wanted to avoid discipline-related problems. Over half of the respondents also cited payment of lower wages, payment of fewer benefits, and the preference of workers for such work (DOLE 1997).

UNIONIZATION AND INDUSTRIAL RELATIONS   That firms admitted to employing contractuals and casuals to "prevent workers from organizing" sheds light on another characteristic of the industry: the low level of unionization, even by Philippine standards. According to the Bureau of Labor Relations/DOLE, out of 367 firms with a workforce of approximately 190,000 there were only 36 registered unions with a total membership of 7,642 in the industry (BLR/DOLE 1999). Those actually covered by a collective bargaining agreement (CBA) are even fewer: there were only four establishments with CBAs that covered only 2,074 workers. Thus only 1 percent of the industry and 1 percent of the workforce are covered by a CBA.

The number of strikes and lockouts in the industry is also low and declining. In 1992 there were a total of thirty-four notices of strikes or lockouts involving some 20,650 workers. By 1996, the number dwindled to

2. Note that 155, or 58.9 percent, of the total sample of 263 electronics firms surveyed reported to have surplus labor (DOLE 1997).

Table 2.8. Strikes and lockouts in the electronics industry, 1992–1996

Year	Number of notices	Workers involved
1992	34	20,653
1993	35	18,457
1994	38	10,457
1995	28	6,445
1996	20	3,280
Total	155	59,092

Source: NCMB cited in BLR/DOLE 1999

twenty notices involving a mere 3,280 workers. This represents an 84 percent decline of workers involved in formal disputes despite a steep rise (116 percent) in the overall number of workers in the industry (see table 2.8) (BLR/DOLE 1999).

The extremely low level of worker organization in the industry, despite the high percentage of regular employees and dominance of MNCs, can be attributed in part to a strong antiunion bias in the industry. One manager stated, "The electronics industry is 'allergic' to unions" while another pronounced that "a semiconductor company's worst nightmare is a union."

But most important, due to stiff competition and just-in-time production chains, the industry is quite vulnerable to any interruptions of production. Collective job actions such as work slowdowns, mass absenteeism, and strikes are viewed as a potentially destabilizing factor that can jeopardize production and delivery schedules. As electronics firms must respond quickly to changes in the market, most firms view unions not as partners or even legitimate worker representatives but as barriers to flexibility and a threat to "management prerogative."

State Regulation of the Labor Market and Labor Relations

> The Japanese are now our biggest investors. But there is only one fear the Japanese have, problems with uninterrupted production. Many locators want a 'guarantee' of no strikes. Without it, they won't locate here.
>
> Director, Department of Trade and Industry
> —Center of Industrial Competitiveness

Because the state has chosen a development strategy based on luring foreign direct investment in electronics—a highly competitive and aggressively antiunion sector—it has been deeply involved in regulating labor relations and the labor market, even as it has "liberalized" the economy. A key strategy to balance labor control and liberalization is devolution. As the government's Medium-Term Comprehensive Employment Plan (1999–2004) states, "Government shall create a more effective and efficient public sector by re-engineering and right-sizing the bureaucracy and forming a more equitable sharing of responsibilities and resources with Local Government Units (LGUs)" (MTCEP 1999, 3). This is clearly in line with what Peck (1996, 72) argues is the continued need for regulation: "The state's role in the labor markets stems from a requirement to secure the necessary conditions for the reproduction of labor-power and, more broadly, of market relations." Wage restraint, labor discipline, and the labor market have thus been key arenas for the state's "roll out" of neoliberal policies since the early 1990s.

One of the first areas for intervention—or strategic nonintervention—has been in the area of wages and their increasing "flexibility." The nation state has long been active in setting minimum wages to keep pace with or run ahead of inflation. But in 1990, just as the trends toward increased liberalization and peso devaluation were taking hold, the setting of minimum wages was devolved from the National Wage Council and congress to regional tripartite wage and productivity boards in each of the sixteen administrative regions in the country to make it more responsive to local conditions. The tripartite councils, which include representatives of government, businesses, and labor, are expected to consider the need for both national and regional "competitiveness" in setting regional wage rates. Specifically, they should take into account "the need for industry dispersal from urban centers to rural areas . . . [and] the absorptive capacity of employers for wage adjustments" (ILS 1997, 18). It should be noted that labor union federations had their greatest strength at the national level; thus devolution of wage setting severely undermined their bargaining leverage and also their power on such tripartite boards and in wage setting generally. Although the boards did periodically increase the minimum wage rates after devolution, the real average minimum wage in the Philippines actually fell, from 83 pesos in 1991 to 73 pesos in 1996 (World Bank 1997). In fact, although the average real wage grew 1.5 percent a year from 1994 to 2002, overall, the real average wage rate in 2002 was still only three quarters of what it was in the early 1980s (Felipe 2004). The decline in real wages, due in large part to devaluation-induced inflation and the weaker bargaining power of labor, was the primary reason unit labor costs in Philippine manufacturing fell more than 60 percent from 1989 to 1997, providing a competitive edge in attracting labor-intensive investment to the Philippines (Felipe 2004).[3]

3. Unit labor costs, a ratio used to compare competitiveness across countries, is defined as the ratio of the wage rate to labor productivity.

A second key area of state intervention has been the disciplining of labor—or put another way—the "freeing" of investors from the constraining demands of labor unions. As noted above, foreign electronics firms put a high premium on labor "flexibility," including the "absence of restriction on management on the shop floor" (McKendrik et al. 2000, 233). As one government report on the changing character of labor-management relations noted, "The policy shift from import-substitution to export-orientation and globalization necessarily changed the rules of conducting business, which in turn, changed the rules between labor and management" (ILS 1997, 23). To this end, the Philippine state moved quickly to stem the tide of union militancy that marked the 1980s and contributed to the stagnation of FDI during that decade. In 1989, the Philippines passed the New Labor Relations Law, the first of many steps to promote a new era of "industrial peace." These laws helped erect new grievance procedures, as well as voluntary and compulsory arbitration mechanisms, all developed to dampen militancy and severely weaken labor's ability to use one of their strongest weapons—the strike (Gonzalez 1999). At the same time, the state only selectively enforces other, strong prolabor laws already on the books. For example, minimum wage laws are routinely ignored: in 1995, of the more than seventy-seven thousand establishments inspected by the Department of Labor and Employment, it was found that nearly 20 percent were violating minimum wage laws (ILS 1997). And as will be evident in the case studies, laws limiting the maximum hours and days worked are also routinely ignored in order to provide employers greater "flexibility."

The government has also created tripartite consultation mechanisms to promote industrial peace. In 1990, the state established the Tripartite Industrial Peace Council with a mandate to formulate recommendations for policies to enhance employment, workers' welfare, and harmony in labor-management relations (ILS 1997). The council can also form subnational and sectoral tripartite councils and has helped develop a Code of Industrial Harmony to support the state goals of stability and sustainable growth. Similarly, with the creation of PEZA in 1995, the state agency set up yet more tripartite bodies—this time at the level of individual zones—to form social pacts for "the enhancement and preservation of industrial peace in the Ecozones" (ILS 1997, 6).

The devolution of labor control to the local level is also evident in the increasing role that the provincial government and local security agencies play in actually enforcing "industrial peace" in and around the EPZs. Many managers interviewed admitted that they prefer locating in economic zones because of the low levels of unionization. Many zones are notorious for their antiunion stances, backed up by local government officials, mayors, and governors who are desperately trying to attract investment and jobs to their areas. This means that union organizers not only contend with the company officials, but also with some local governments. In Cavite, for example, the former governor, declaring the entire province a "peace and productivity zone," set up the Industrial Security Action Group, a special intelligence-gathering and police force to patrol both private and public zones, and brutally enforced a "no union, no strike" policy in the province for most of his two stints as governor, from 1979 to 1986, and from 1988 to 1995. The antiunion strategies in and around the zones will be examined in much closer detail in later chapters.

In addition to increased consultations, the state has promoted alternatives to trade unionism to move further away from the traditional "adversarial" union model. In particular, the state has encouraged the development of labor-management committees (LMCs) or cooperation schemes at the plant level, which it considers "a necessary condition for survival and growth in a highly competitive market" (ILS 1997, 57). Indeed, one of the lead agencies providing training on the creation of LMCs is not the Department of Labor and Employment, which has the overall mandate to protect workers, but the Department of Trade and Industry's Center for Industrial Competitiveness—a subagency concerned with increasing foreign investment. While the state has touted the economic benefits of cooperation and participation in joint productivity boards or quality circles, and has even made LMCs mandatory for drawing on certain government incentives, unions have resisted these schemes in large part because they view them as an attempt to substitute, rather than enhance, organized labor's voice in negotiations.

Part of the state's rollout of new policies to improve the attractiveness of the Philippines to advanced manufacturers is the restructuring of its technical education programs in partnership with both local nonstate actors and foreign investors. In 1994 the state established the Technical Education and Skills Development Authority (TESDA) to regulate training and vocational education programs and better tailor the development of Philippine labor to market needs. While TESDA directly operates a network of fifty-nine regional and provincial training centers and oversees other public training institutions, 80 percent of the students enrolled in technical and vocational courses are in privately run organizations such as Dual Tech, the Don Bosco Foundation, or the Meralco Foundation. One strategy of the government has been to coordinate and set up apprenticeship programs between large employers and training institutions in order to provide firms with specifically skilled workers. This strategy has been particularly popular with electronics firms, which have faced a bottleneck in terms of finding enough qualified technicians. Often, these firms donate the machinery and are allowed to shape the curriculum to fit their needs. Trainees then spend 70 percent of their thirty- to thirty-six-month training period working in the firm and spend the rest of their time at the training center. Firms pay only 75 percent of the minimum wage to these trainees and are under no obligation to hire them after training (though many do hire from among their apprentices). Trainees retain 30 percent of their pay, with the remainder being paid to TESDA or the private training institute (Esguerra et al. 2001). Such training programs allow firms to develop a pool of highly qualified workers with industry- and often firm-specific skills, while at the same time allowing firms to screen potential workers and pay well below market rates during the long apprenticeship period.

Finally, the Philippine state has also gotten more deeply and directly involved in the labor market and worker recruitment at the local level. In 1992, the Department of Labor and Employment set up the Public Employment Service Office (PESO) program in order to devolve employment services to the lowest administrative level. The program works directly with mayors and local government units (LGUs) to coordinate job placement for large and small businesses, conduct job fairs for individual companies and EPZs, and to register applicants into the National Manpower and Registry System to pool labor market information. The PESO Act of 1999 enabled the establishment of PESO offices in every province, key city, and urban municipality, and as of 2000 there are at least 1,850 offices (Esguerra et al. 2001). While the offices do provide local applicants with much needed labor market information and access to jobs, because they are run through the offices of local elected officials, they can often be used a tool for patronage and political screening—with mayors and other officials using the jobs as a way to reward supporters or blacklist union organizers.

Conclusion

This chapter has focused on the changing demands in the global electronics industry, the changing character of the Philippine state and its development strategies, and the Philippine labor market. Beyond describing these factors individually, I have tried to develop a more dynamic vision of their interconnected character and relationship to processes of globalization.

In terms of the electronics industry, the forces of global competition and technological change have led to a global industry structure that appears to be contradictory: dispersion with concentration (Ernst 2002). While production is increasingly spread across the globe, rapid changes in technology have created exploding investment costs and thus very steep barriers to entry in the high-value-added sectors, such as research and development, wafer fabrication, key component manufacturing, and even supply-chain management. Consequently, the global industry is dominated by MNCs that coordinate dense but dispersed production networks. These firms, facing new forms of competition and production, have shifted their locational strategies. Although value chains have long been spatially fragmented, each node or cluster must also now be both more flexible and more tightly integrated within production and supplier networks. Leading multinational firms must now look beyond pools of cheap labor and instead seek to combine particular labor supplies with the proper regulatory conditions that promote stability and "business friendly" clustering or agglomeration.

This, of course, has significant implications for those localities that hope to draw such investments. The Philippines has been partly successful, drawing a huge influx of investments and producing record-breaking exports. Nevertheless, the domestic industry remains structurally stagnant: dominated by large multinational firms, dependent on imports, and highly concentrated in low-value-added semiconductor assembly. This import dependence has led to a "missing middle," or dearth of small- and medium-sized Filipino supplier firms. The lack of local suppliers severely limits technology spillover, local value-added, and thus the benefits of attracting MNCs into the country. In fact, although now more capital intensive and "high tech," the industry in the Philippines remains in essentially the same position along the value chain as it did on entry: at the bottom. However, even at the bottom, the Philippines in the long run cannot compete with the likes of China, now the global benchmark standard for low-wage assembly manufacturing.

Why has the Philippines succeeded in drawing international investment but failed to upgrade the domestic industry? The answers lie in large part with the shifting capacities of the Philippine government and its further retreat from a developmental state strategy. Even when "strong," the Philippine state has more often been "predatory" than developmental, more overrun than autonomous, and its industrial policies far more incoherent or contradictory than transformative. Driven by the tides of global neoliberalism that have recast national sovereignty, the Philippine government has crystallized into a regulatory state focused on providing to the global market unique local resources and conditions at the manageable scale of export processing zones. In essence, the Philippines' development strategy is really a labor market strategy, built around what it considers its comparative advantage: cheap, skilled, English-speaking labor. As I have shown, the Philippine labor force offers a number of unique "fixities"—an abundance of educated production workers and technical and managerial talent awash in high levels of un- and underemployment at all education levels. But as Henderson notes, "Investment [tapping low cost technical and engineering labor] does not arise merely because of the existence of such labor supplies. . . . the host economy must be capable of delivering the social arrangements and managerial competence necessary for high productivity and the production of high quality products" (Henderson 1994, 117). What matters then is social and political regulation of production and the labor market. In the Philippines, this strategy is embodied in the state's revamping of the EPZ program, and its investment policies, which combine an overly generous incentive package, poor support for local firms, and the selective enforcement of labor laws.

This chapter began with describing the production demands that drive subsidiary-level business strategies, which in turn influence where and how firms organize production. Of course, these demand factors alone by no means determine how work is organized locally. However, they do set limits on local decisions and, along with local regulatory structures and labor markets, set the parameters within which work itself is negotiated and struggled over. But the complexity of competition and devolution means that much of the actual regulation, or what Brenner and Theodore (2002) call "actually existing neoliberalism," and what I call here local labor control regimes, happens at the subnational level in both formal and informal ways. In the next few chapters I move from the macro-level to the factory floors, local communities, and personal lives of workers to investigate just how such complexities are worked out and at what cost.

3   High Tech, High Performance?: The False Promise of Empowerment

ADVANCED ELECTRONICS multinationals face a contradictory and often brutal market. The demands for cheap, high-grade, and innovative products mean that firms must sink substantial resources into flexible but high-volume production, churning out high-quality yet inexpensive products that meet customer specifications. Such "mass customization" means that firms cannot simply rely on arm-twisting cheap labor. Rather, they need more committed and productive workers who can help stabilize production and "marginalize uncertainty" (Delbridge 1998). But how do firms actually restructure work to achieve both mass customization and a more productive and stable workforce? Is there—as some scholars argue—an emerging "one-best-way," which bundles new technologies with flatter hierarchies and worker empowerment (Kalleberg 2001)? And what do these new strategies mean for front-line workers, most of whom are women?

In this chapter, I focus on the factory floors of four high-tech firms and compare their production strategies against three hallmarks of high performance: flexibility, quality, and productivity (Appelbaum et al. 2000). I then turn to how work is actually organized and how firms elicit the three key elements of worker commitment, namely: loyalty, effort, and attachment. The cases show that rather than following a single model, the firms pursue a wide variety of production and commitment strategies, using different combinations of "hard" engineering-intensive and "soft" human resource (HR) approaches that have quite different consequences for workers. I then explore this variation, focusing on three key mechanisms that help explain divergent firm strategies and outcomes, including firm nationality, the character of the firm's product and production, and the balance of power between labor and management. As I will show, while the firms take different paths to achieving flexibility and show some variation in worker outcomes, their work regimes nevertheless all in some way expand management control over pro-

duction, *narrow* worker discretion, and *widen* the divide between production operators and technical and engineering labor. None, then, substantially
delivers on the promise of workplace "empowerment."

I also examine in this chapter the impact of restructuring on women, and
how gender is constituted through the division of labor and the organization
of flexible work (Wright 2001; Salzinger 2003). In an industry in which
women have dominated production, work reorganization has been held up
as an avenue for increasing women's employment and socio-economic mobility (Kuruvilla 1996; Yun 1995). Yet as the cases will demonstrate, while
restructuring can *potentially* improve women's occupational opportunities
and bargaining power, the gendering of new technologies and the widening
gap between manual and technical labor may in fact increase the divide between "feminized" assemblers and "masculinized" technical experts. And as
manufacturing becomes increasingly automated, downsizing and a "masculinization" of production labor may create a more deeply divided labor
force in which women workers witness—but have no access to—more
skilled and stable production jobs.

It is worth reiterating here that the labor processes and work regimes profiled in this chapter are but one aspect of flexible accumulation, which critically depends on a wider political apparatus beyond the factory gates that
exploits and extends existing social divisions in the Philippines and in local
labor markets. Thus while the logic of flexible accumulation in each of the
case studies begins with particular "internal" plant strategies, their con-
comitant "external" interventions—examined in the next chapter—work in
tandem to help secure worker commitment, and uninterrupted production.

Allied-Power and the Despotic Work Regime

> I don't think they look at us as workers. It's like we're brooms.
> When they need us, that's OK, they take us, but if they're done using
> us, they just dump us in a corner.
> Production worker, Allied, Ltd.

The Allied-Power group represents the despotic work regime, characterized
here by strong direct control by management, low wages, harsh working
conditions, frequent layoffs, and active suppression of unionism. It is the
work regime that resembles most closely the popular image of "sweatshops"
in Philippine export processing zones (EPZs) that has persisted since the
1970s (Ohara 1977; Aldana 1989; Chant and McIlwaine 1995; Klein 2000).
Of the four work regimes, it is also the least flexible and most traditional in
its organization of production.

Power Tech, the first of the two firms in the Allied-Power group, is located at the Cavite Economic Zone (CEZ) in Rosario, Cavite, some ten miles southwest of Metro Manila.[1] It is a wholly owned subsidiary of Power Tech Korea, a medium-sized multinational that produces low-end consumer electronics. Power Tech is a subcontractor for large office phone suppliers but also produces its own brand of radios and phones sold mainly in U.S. discount stores such as Wal-Mart. Because technology is simple and market entry barriers low, Power Tech competes based almost exclusively on price and faces particularly strong competition from producers in China.

Opened in 1989, Power Tech's $10 million Philippine plant produces cordless phones, printed circuit boards (PCBs), CB radios, corded telephones, and cable TV converters. Although the production area is generally clean, the low-tech nature of the products does not require expensive, dust-free "Clean Rooms" as in the other more high-tech firms discussed later in the chapter. Production remains labor intensive, but Power Tech has begun to purchase more automated machinery. Power Tech has a total of 935 employees, of which 82 percent are factory workers paid on average 215 pesos per day or about P15 above the regional minimum wage.[2] However, due to a lack of orders over the research period, production was far less than capacity and 80 percent of the workforce was working less than forty hours per week. This meant that average monthly take-home pay was only P4,400 or US$116 and the firm's monthly output was $2,349,236, or $2,419 per employee.

The other company discussed here is Allied, Ltd. Allied began operations in 1991 as a subcontractor to Power Tech, producing electronic transformers and adaptors for cordless and cellular phones along one assembly line within Power Tech's factory. In early 1995, Allied constructed its own building within the Cavite Economic Zone and began separate production. Allied also began producing for other, mostly Korean, firms and even managed to land several subcontracting jobs with large Japanese firms.

Allied operations are very modest: production takes place in a cramped, cinderblock building in warehouselike conditions. Although the building has several small air-conditioning units, workers complain that these are wholly inadequate, particularly for workers in the soldering section. At the time of research, Allied had total of 306 employees, with factory workers officially paid the regional minimum wage of P200 per day. But because of a severe slowdown, most employees worked only one to three days a week, so monthly wages were only about P3,000 or US$75 (according to the Ecu-

---

1. Originally named the Cavite Export Processing Zone (CEPZ), the zone was renamed the Cavite Economic Zone or CEZ in 1995, following a change in federal law recasting the Export Processing Zone Authority (EPZA) as the new Philippine Economic Zone Authority (PEZA). The term used throughout will correspond with the period being referred to.
2. At the time of the research in 1999, the exchange rate was approximately 38 pesos to US$1.

menical Institute for Labor Education and Research (EILER), a "living wage" in the region for an average Filipino family of six is P424 per day or P10,504 per month).

Total monthly firm output was valued at $56,553, or $185 per employee. Comparing this output with the $25,799 per employee output at Integrated Production (discussed below) or the average monthly earnings at another firm in the study, Discrete Manufacturing (P10,983 per employee), one gets a sense of the enormous range of output and wages in the Philippine electronics industry.

Flexibility, Quality, and Productivity Strategies

FLEXIBILITY   Production flexibility can be defined as "the ability to quickly, efficiently, and continuously introduce changes in product and process" (Deyo 1997, 214). While flexibility is crucial at all market levels, Power Tech and Allied Ltd. produce for the low-end of the global market, where price is key and orders fluctuate widely. The firms respond to such market demands almost entirely through external or numerical flexibility schemes. In contrast to internal or functional flexibility, in which firms adjust the *process* of production through such schemes as line reorganization or use of work teams, numerical flexibility involves adjusting the *amount* of labor used in production. For example, when demand for its products spiked in 1995, Power Tech merely intensified work, resorting to speedups and heavy overtime. At that time, they employed more than 3,000 workers and operated three shifts a day, six to seven days per week. They also contracted out work and cross-posted workers with affiliated firms. Conversely, as demand fell in 1997 due to quality problems, the Asian Financial Crisis, and increased automation, the firm shed more than two-thirds of its workforce and made liberal use of slowdowns. By 1999, Power Tech was down to less than 750 production workers and the firm operated only one shift, five days per week. Similarly, Allied, at its peak of production just prior to the Asian Financial Crisis had 750 production workers and output was fourteen times higher than in 1999.

In terms of technology and work organization, both firms rely on assembly line layouts, adhere to rigid organizational hierarchies, and use mainly older technologies. At Allied, one engineer noted that the firm only gets machines that are no longer used in Korea. For example, in the winding section, where wire for transformers is wound onto spools, one operator explained, "We have new machines, but still the same old technology. The old one [used until 1995] was foot-pedal powered. Now it's both push-button and foot-pedal operated. If your foot gets tired, you can switch to push-button." Another worker complained of the lack of safety equipment:

> In winding, your hands get sweaty and you have only one pair of cotton gloves for a week. And the counter: if you have no gloves, it's hot and it hurts your fingers. The counter winds the wires, but you have to handle the wire to support

it while it winds, so it gets hot. . . . If you don't have gloves, you're done for. The heat gets so high, you get burned.

The other major area, the soldering section, is also entirely manual and workers complained of heat and exposure to fumes. An operator explained, "In soldering, no masks, no gloves. Well, they do have gloves, but you have to ask for them and when you ask, they say they don't have anymore. So we have to endure the old ones or none at all."

QUALITY   Since both Power Tech and Allied compete primarily through price, neither has the sophisticated quality programs found in the other three firms discussed in this book. At Allied, product quality is particularly low. According to Jerry,[3] an engineer who resigned due to lack of work and poor conditions, the quality of the imported Korean raw materials were often poor, contributing to low-quality products. Reject rates are also many times higher than in the other three firms discussed later. Jerry, contrasting Allied to other firms, said, "Allied is interested in quantity, not quality. Our reject rate was 2 percent, whereas the Japanese are at 'point something' percent. They [Allied management] won't stop production just for some little quality problem."

Similarly, Lisa, who began as an operator and later moved to line-leader and quality assurance, explained:

There are a lot of rejects that get through. For example, from a thousand pieces, we sample only two hundred. So you can't see all the rejects. The problem is they don't check the process. . . . They just keep producing, just packing them in. But when it's finished, the winding, they complain there are lots of rejects. But I say, "it's your fault, not ours. You didn't inspect [the machines]. You need to assign another inspector." Or, if we are not doing something, we could be made to inspect. They could place us there. But they don't. And there are cases [of products] that they've already sealed, but they have lots of rejects. Not only one hundred but two hundred [rejects] in one box. A lot of waste.

While the other three firms in this book view international quality certifications such as ISO 9002 and ISO 14000 as absolutely essential to global competitiveness, Allied has no international quality certification, no quality programs, and very little training.[4] Again Jerry:

---

3. All names of the factory workers used in this book are pseudonyms.
4. ISO 9002 is a set of quality standards for all aspects of production and service developed by the International Standardization Organization in Geneva, Switzerland, in 1987. ISO 9002 certification requires a company to follow strict standards and procedures for measuring quality in each part in production. ISO 1400 is a similar certification based on implementing an environmental management system.

The trend in the industry is to be ISO certified. But the [Allied management] is not willing to try because they say it's too expensive and it would be too difficult for them to get certified. They don't see the need yet for ISO 9000. But we had buyers from Japan. They had higher quality standards and their auditors rejected our products.

While Allied shows little interest in quality control, Power Tech was attempting to move into higher quality production, mainly by introducing more automation. The firm purchased a number of new machines such as a high-speed chip shooter for automated surface mounting on printed circuit boards (PCBs), automated insertion machines, and multifunction pick-and-place machines. It was hoped that these would dramatically improve the production quality by reducing the amount of operator handling and improper assembly.

In addition to the new machines, Power Tech has also become ISO 9002 certified. The certification means that all operators must be trained and certified at their workstations and must keep strict quality records. Advocates of ISO certification have argued that it can improve worker skills and put a check on management's arbitrary power because all workers are trained and management must follow strict procedures. However, workers interviewed said that training and ISO certification is not linked to increased discretion or pay. While skills might be enhanced, they are machine- or station-specific and therefore not transportable to other jobs. In fact, many operators complained that ISO certification simply added more paperwork to their already fast-paced jobs and felt that the documentation would only be used to evaluate their individual performance. ISO can also *reduce* the number of worker skills, since some firms are reluctant to spend both time and resources on certifying workers on multiple processes. For example, Power Tech and Integrated Production discussed later, have actually cut down on the amount of multiskilling, opting instead to have workers dedicated to a single machine or process.

Clearly, Power Tech, unlike Allied, understood that quality was a key element in remaining competitive. However, this new attention to quality has not positively affected workers. First, according to the HR manager, operators still only go through one week of initial training and work only one or two low-skilled stations. Second, the firm achieved higher quality primarily through automation, which led directly to the retrenchment of more than half of its workers. While such a shift to "leaner" production is sometimes accompanied by an increase in positive incentives for remaining workers, this was not the case at Power Tech. Thus, in terms of quality, both firms' strategies stand in marked contrast to the elaborate quality schemes, such as total quality management (TQM), quality teams, or automated statistical process control (SPC), that are integral to manufacturing at the other firms examined in this book.

PRODUCTIVITY   Low-end producers must obviously maximize productivity to keep unit costs down. As is consistent with a despotic work regime, both Power Tech and Allied plainly treat labor as a variable cost to be minimized rather than a resource to developed (MacDuffie 1995). During periods of high demand, both firms utilized forced overtime and forced work on rest days to keep production at full tilt. But in slack times, they quickly resorted to retrenchments.

Still, even with slack demand, management at Allied often ratchets targets, and production quotas are a constant point of struggle. One worker clarified:

> In winding, the quota is a thousand pieces per day, it's individual. It's not difficult. . . . but there's no incentive for you to go over your quota. Even if you work fast, it amounts to nothing. We just get tired. We just maintain one thousand, until it's 4:30.
>
> See, our quota was eight hundred. But when we reached that, they raised it to one thousand. Then they raised it again to twelve hundred. But no, we stayed at one thousand. We're able to do twelve hundred but we stayed at a thousand.

The high quotas also led to increasing animosity between managers and line leaders who felt loyal to "their" operators and thus tried to negotiate lower targets. Lisa, a former line leader, explained: "I was supposed to assign [the positions] to my co-workers, but [the supervisor] would intervene and give us a quota. I'd tell her, 'My people can't handle that! Couldn't the quota be just a little less because they can't do six thousand in eight hours.'"

The management at Power Tech took a similar approach. The HR manager explained, "When production was high, there was no tolerance for taking unnecessary leave." However, as demand fell, the firm simply cut back on production, laid workers off and cut workers' hours: "Now, if they want to go on leave, they are allowed, with much ease." The HR manager also claimed that a major reason demand fell in the first place was a unionization drive, begun in January 1997. According to the HR manager, "I believe the lack of orders and the union drive are related. The union drive led to a slowdown by workers, which then led to low productivity and low-quality work. This discouraged buyers, so their orders dropped and so production had to drop too." The manager, however, neglected to point out that the workers' organizing attempts were in response to the first major retrenchments in 1996. Since the failure of the unionization drive, the manager noted that productivity was on the rise because, he said, "workers are listening to what the company explains to them again and they are much less resistant to what you tell them."

Control, Human Resource, and Labor Relations Strategies

DISCIPLINE, PAY, AND BENEFITS   With production organized along an assembly line, the pace of the line and production quotas help discipline workers and keep them on task. However, production levels often fluctuate and management must often "convince" workers to meet company demands. Yet in the case of both Power Tech and Allied, management shows little concern for promoting worker commitment and instead organizes employment relations in a traditionally top-down way, wielding a heavy disciplinary stick.

At Allied, discipline revolves around the strict enforcement of company rules. Workers reported that management bars operators from leaving their stations during production, even to go to the bathroom. This has prompted some workers to bring plastic bags to their workstations into which they urinate in emergencies. Discipline is also visible and strict. One worker recounted two incidents when she and another were given disciplinary actions (DA) in the form of badges, which they were then forced to wear:

> Cecile, she put in place the rules and regulations, she's really strict. She's the one that gave me the badge. She gave me a yellow one. It has a meaning: you get the yellow badge if you argue with them a lot. Blue is for an obedient one. Red is if you're really hardheaded and defiant.
> I'd arrived five minutes early, 5:55 [A.M.]. Then sounded "ding-dong" [shift start signal] over and over. My co-workers were all new and they were in front of the mirror. So I hushed them and said, "Hey, there's the 'ding-dong,' you have to go in now." I still had to eat. Well, they still didn't go in and Cecile's "ding-dong" continued until it was six [A.M.]. Finally, they started walking in. Cecile saw me and I was at the end of the line. "Hey, Garcia! Come here, come here!" she said. "You're a regular, you've been here a long time and you wouldn't even reprimand your co-workers? What were you doing?" Well I couldn't tell her I was eating, that would've made it even worse. So I said, "I told them to go on in," but she still gave me the yellow badge.

In a separate incident she explained how her co-worker got a DA just for singing: "They [the supervisors] are really strict and so people hate them. Like when Debbie sang, when she was pregnant. Julia [another supervisor] passed by, and it seems the song got louder. So they gave her the badge."

The "badges" were clearly meant to humiliate workers and also affected their yearly appraisals used for determining raises and promotions. Receiving several DAs, particularly "red badges," constituted grounds for "justifiable termination." The strict management hierarchy within the firm often strained the relations between workers, supervisors, and managers. Lisa recalled why she decided not to remain a line leader: "We started having problems and the Koreans always got sore with us. When [the supervisor] was scolded by the Koreans, she took it out on us."

The authoritarian management style was not, however, compensated by more positive incentives to promote commitment, as is the case at Storage Ltd. discussed later. Instead, Allied management minimized all costs and offered few positive rewards. In fact, the company opted to put remaining workers on rotation and forced vacation instead of officially retrenching them. The company schedules workers as little as three days a month, in hopes that they will simply quit, thus relieving the firm of the legal obligation to provide severance pay. This has meant extremely low take-home pay for workers. One worker commented: "Sometimes, you only get 500 pesos because you only work a few days a week and there are deductions. And there's the boarding house. Now, I commute from Manila because I can't afford the boarding: the meals, the rent, transportation fares. I'm really short." Another worker lamented, "On one pay stub, for 15 days, after all the deductions, I was left with two pesos."

The "deductions" were another source of contention. By law, employers must contribute and also deduct employee contributions to the national Social Security System (SSS), Medicare, and a social fund (PAG-IBIG), against which workers can borrow money. Workers claimed that Allied has been deducting their contributions but has not remitted the money to the government. This was discovered when workers tried to borrow from the emergency loan fund and found the contribution behind by about one year. The firm also stopped providing other legally mandated benefits, like paid vacation leave, at the same time they cut off incentive pay. In addition, there is no transportation allowance, no housing allowance, no shuttle, and no rice allowance. Job security is quite tenuous, and in fact, workers *want* to be retrenched so they can at least collect their separation pay, remaining there only in hopes that they will get this final compensation.

At Power Tech, the situation is marginally better. Workers tried to organize a union after initial retrenchments in 1996, but eventually failed. After the organizing attempts, direct control was reinforced through close supervision. For example, there is now one supervisor for every seven operators. As the HR manager stated, "The union storm has settled. They realized the effects of what they started; they saw that it only led to lost jobs for their friends and relatives. So they stopped. Now they are back to harness."

But the company has also introduced some positive incentives since the union drive. They do not employ any temporary or contractual workers, and the pay rate is slightly above minimum wage. The firm claims to provide all legally mandated benefits, such as SSS, PAG-IBIG, Medicare, life insurance, maternity leave, and a yearly thirteenth month pay bonus as well as a meal allowance. However they do not pay any retrenchment benefits. In fact, the HR manager claimed that all firms in the CEZ, although under legal obligation, fought paying severance pay or simply ignored the law. And Power Tech, like Allied, has slowed production and rotates workers to take advantage of a "no work-no pay" policy; consequently monthly take-home pay is well below even minimum wage.

Power Tech has thus made some changes to their disciplinary system, relying less heavily on negative sanctions and at least complying with the Philippine Labor Code on many provisions. However, these changes must be understood in connection with the firm's union-busting strategies and large-scale retrenchments that also had a chilling, disciplinary effect on remaining workers.

LABOR RELATIONS AND UNION BUSTING   Workers at both companies have attempted to unionize primarily because of the layoffs, harsh working conditions, low pay, and nonpayment of benefits. The management at both firms has responded aggressively to the unionization drives with violence, threats, bribes, forced shutdowns, retrenchments, and the shifting of production to nonunionized firms.

At Power Tech, workers began organizing around the first wave of retrenchments in 1996, targeting more than a thousand workers.[5] According to the HR manager, the "union storm" that swept into Power Tech was entirely instigated by "outside organizers trying to penetrate CEPZ [CEZ] because there were no unions before. They saw it as a training for themselves, challenging themselves to organize the unorganized. They lured [Power Tech] workers into joining an outside union."

In fact, after several months of clandestine contacts with several union federations, workers opted to organize as an independent union, unaffiliated to a major labor federation. Rather, the workers contacted a local Catholic priest who had been instrumental in organizing several Christian unions affiliated with his parish beginning in 1994.

During the registration drive in January 1997, in which the union needed 20 percent of the workforce as members, the management became aware of the organizing. After the union successfully registered with the Department of Labor and Employment (DOLE), the company announced a voluntary retirement program, hoping to entice the 1,150 workers targeted for retrenchment. While about 650 workers accepted the company's buyout offer, the remaining workers did not, and vowed to fight for their jobs. When workers staged a "noise barrage" inside the factory, banging pot lids, honking horns, and yelling at their workstations, management retaliated, accusing the workers of staging an illegal work stoppage. The firm first suspended, then fired, twenty-five union leaders and active members.

The firm's other tactic was to shift its production. Power Tech had established another firm in the CEZ, Ultra Electronics, in early 1996, at about the

5. This was also the period when the former governor of the province, Juanito Remulla, lost the governorship and labor organizers for the first time began to operate aboveground in Cavite. The discussion of the political context for labor organizing around the Cavite Export Zone will be discussed in greater detail in chapter 4.

same time as retrenchment began. Several of Ultra's production lines came directly from Power Tech and the majority of Ultra workers were transferred from Power Tech. These transferred workers included the remaining leaders and core members of the union.

Power Tech's strategy, then, was to crush the strike and simply "behead" the fledgling union by either retrenching, paying off, or transferring its leadership and core members to intimidate the remaining workforce. For the firm, the strategy worked: there is no longer a union at Power Tech. The HR manager spun it this way: "There is no union now because their registration [with the DOLE] did not push through. One by one, the general members abandoned the ship. The officers abandoned the workers, so workers abandoned the union. They learned their lesson, so I don't think there'll be another union here."

Allied, as a member of the Power Tech group, shares the same "HR consultant" as Power Tech and Ultra Electronics and thus the same labor relations strategies. Workers at Allied began organizing themselves into a union following a drastic reduction of the workforce from 750 to 250 in late 1997. Using a similar strategy that was deployed at Power Tech and Ultra Electronics, the Allied management vigorously fought the formation of a union, first by challenging the union's registration and certification election at the Department of Labor and Employment, then through intimidation and bribery of the union leadership. And, as in the case of Power Tech, the Allied management was initially successful in dismantling the union. However, Allied workers have been able to regroup and reform their union and have managed to gain formal recognition from the Allied management, despite ongoing threats that the firm would shut down and move production to China. The actions of the Allied workers in trying to organize will be taken up in more detail in chapter 5.

Thus both Power Tech and Allied chose straightforward union suppression and union-busting strategies. As will be discussed in chapter 4, these tactics were available to the firms in large part because they are backed up by the coercive muscle of the zone police and hired goons who enforce the "no union, no strike" climate within the CEZ and within the province. Workers at the two firms were not successful in appealing to either the zone administration or the provincial Department of Labor and Employment (DOLE) for assistance, thus forcing them to appeal to the DOLE at the national level, where actions are slow and where their decisions may not be enforceable locally. The combination of the firms' impunity in the face of such illegal actions and the real, credible threat of retrenchment without pay have had a strong disciplinary effect on remaining workers and has indeed help bring them "back to harness."

Despite their "success" in busting the fledgling unions, both Power Tech and Allied were struggling to stay alive. Allied had lost important Japanese customers due to poor product quality, resulting in irregular work and lay-offs. Power Tech had similarly cut back production and reduced its work-force by three quarters. Labor relations at both firms were adversarial, and at Allied, deteriorating working conditions further demoralized existing workers, making it harder to maintain quality or fill existing orders. The recent problems at both firms and the rise in union organizing show that a despotic work regime, which uses few positive incentives and fosters no worker commitment, can generate instability and open conflict. The failure of the firms to contain such instability underscores the fact that shop floor coercion has not been able to provide what many more advanced manufacturing firms require most: conditions for stable yet flexible production.

Storage Ltd. and the Panoptic Work Regime

> For most operators, they have no control. They are treated like ro-bots; told what to do, when, where and how. Every movement is controlled. Even outside the production area, they are controlled; they are told where to go, when to eat. And everything is timed, even going to the bathroom.
>
> Human Resource officer

Storage Ltd., which represents the panoptic work regime, shows that high-tech, high-quality production can take place in the Philippines, and also that worker commitment is crucial to the success of flexible manufacturing (Appelbaum et al. 2000). But the case study also reveals that commitment, productivity, and high quality do not necessarily require greater worker involvement or the reduction of workplace hierarchies and can be quite compatible with strict worker control. Indeed, the case confirms other studies showing that increased technological intensity may lead to a deeper technical and gendered divisions of labor (Kelker and Nathan 2002).

Storage Ltd., a wholly owned subsidiary of a leading Japanese electronics conglomerate, began operations in 1996, assembling and testing high-end computer hard disk drives (HDDs) and magneto-restive (MR) heads for reading stored data. Like most HDD producers, the company does not sell its product directly to consumers, but instead sells its drives to other manu-factures for use in their name-brand products. In the HDD market, profit margins are thin and early—so ramp-up speed to volume production is crit-ical (McKendrick et al. 2000). Production schedules are tight and arranged with suppliers and customers on a just-in-time basis. Final assembly is a cru-cial stage, particularly because of its potential for bottlenecks. For example, turnaround time at the silicon wafer fabrication stage may take two months, while key component assembly, the next stage, takes one and a half weeks. However, the final stage, assembly and testing, may have a turnaround time of only one and a half days. Finally, because of rapid technological change, capital equipment depreciates quickly. The production imperatives at this stage, then, are to maximize equipment productivity, minimize labor costs, and avoid disruptions and delays.

The sprawling manufacturing plant is located in a private or Special Economic Zone (SEZ) just off the new expressway in a formerly agricultural area in the province of Laguna, about forty kilometers south of the capital, Manila. It has three main, interconnected buildings occupying a total of 8.4 hectares, or more than ten times the size of Allied's plant. Unlike Allied-Power (and more like Integrated Production to be discussed below), Storage Ltd. required a massive initial capitalization of $124 million, or more than twelve times that of Allied.

Storage Ltd.'s high-end disk drives and MR heads, like integrated circuit semiconductor chips, are engineering-intensive products and extremely sensitive to dust and voltage fluctuation. Assembly manufacturing thus requires virtually dust- and static-free "Clean Room" production areas and expensive automated equipment, arranged along computerized assembly lines. However, many elements in the production process have proven difficult to automate. Storage Ltd. thus employs more than nine thousand workers, 87 percent of whom are shop floor operators.

At the time of research, production was booming in part because the firm had just brought into mass production their latest high-end drives and was trying to exploit its "first-mover" advantage, when it could charge a premium for its products. During this critical but fleeting period the firm was pushing to produce as many HDDs as possible. Storage Ltd. produces 1.5 million disk drives and nearly 1 million MR heads per month. According to company data, Storage Ltd. paid its direct labor an average of P6,631 or US$174.50 per month. In one sample month, Storage Ltd.'s exports were valued at US$75,000,864, or an output of $8,333 per employee.

Flexibility, Quality, and Productivity Strategies

FLEXIBILITY   Storage Ltd.'s flexibility strategy centers on keeping pace with extremely short product cycles and customized orders while still maintaining high volume and quality. The firm upgrades its two classes of HDDs as often as every six months, requiring frequent design, material, and production line changes. Flexibility is maintained mainly through the use of general purpose, programmable machines. Much of the flexibility in production is designed and developed at Storage Ltd.'s parallel facility in Japan, while continuous improvements are implemented simultaneously in the Philippines. Industrial and process engineers conduct time-and-motion studies and balance assembly lines, while senior production engineers are responsible for actually making output targets and keeping processes running smoothly. As one senior production engineer put it, "Even a single minute is important here. In a semi- and fully-automated environment like this one, if even one machine breaks down, it stops everything." A small army of technicians, nearly all of whom are men, is the engineers' most valued ally, conducting preventative maintenance and solving small technical problems.

This is not to say that operators are completely uninvolved or "deskilled." Given the frequent assembly line changes, workers must be trained to adjust to customization. They must also know enough to recognize when their machines are malfunctioning. For this and other quality reasons discussed below, the company must invest in its own workers: it hires no contractuals and each worker goes through three months of technical training to be certified on three different machines or work stations. This type of labor flexibility, know as multiskilling, allows management to shift workers along and between production lines as demand and production warrants. But at Storage Ltd., workers have no say over flexibility, and multiskilling is used mainly to fill in for absent workers, who are often missing because of the heavy overtime and lack of rest days. In effect, multiskilling ensures a reserve workforce internal to the firm.

Thus the main actors directly engaged in production flexibility and maintenance are engineers and technicians. If a machine breaks down, the operator is explicitly forbidden to try to repair it. Instead, she must stop work and report to the process engineer, who in turn dispatches a technician to the machine while reporting the problem to the production manager.

QUALITY Unlike the low-tech products and processes at Allied-Power, high-quality production at Storage Ltd. is absolutely essential and the firm, like many other high-tech manufacturers, has embarked on a total quality management (TQM) strategy.[6] Storage Ltd.'s TQM system focuses on three areas: continuous computerized monitoring, standardizing work procedures through ISO 9002, and its High Reliability employee monitoring and compliance program.

Like Storage Ltd.'s flexibility program, TQM at Storage Ltd. is primarily engineer driven. The main TQM tool is statistical process control (SPC).

---

6. Total quality management (TQM) systems vary, but most focus on continuously improving quality—in product, process, and people—as the basis for global competitiveness. The main goals of TQM include reducing product and process variation; surpassing customer expectations for price, quality, and delivery time; and reducing "slack time" in design, production, and administration (Hackman and Wageman 1995). Common TQM tools are the systematic quantification and measurement of variation through statistical process control, employee suggestions systems, and off-line quality teams.

Storage Ltd. employs more than thirty computer engineers and technicians to manage their continuous monitoring system (CMS). Operators at each section scan in the bar-coded components as they move through assembly so that individual disk drives can be tracked and any mistakes or "mis-ops" (mis-operations) can be traced to the individual station. In this way, the computerized assembly line acts as a powerful tool for employee surveillance, collecting productivity and quality data continuously on practically every worker. CMS is also used to update the ubiquitous "performance boards" hanging over each section that track productivity and quality and pinpoint unproductive workers.

The second core TQM tool is the ISO 9002 quality system. As mentioned earlier, ISO 9002 is an international quality certification requiring strict procedures for measuring quality. For workers, ISO 9002 requires individual certification to handle particular machines. The company's flexibility goals, coupled with ISO stipulations, mean that workers must have three months of training before becoming optimally efficient. This required investment over a longer period has led the firm away from hiring temporary labor. As the HR director noted, "We hire no contractuals, which can be good and bad. Contractuals are better in the short-term because you can hire and fire them. But in the long-term, they are more costly because they have no loyalty." For workers, certification does give them some leverage vis-à-vis management, since they are more difficult to replace. However, they must continuously document their work, and reporting often means added pressure and an additional self-monitoring layer on top of an already pervasive surveillance system.

Finally, Storage Ltd. uses its High Reliability program to get workers to conform to its overall TQM program. As the firm's own promotional materials state, "The High Reliability programs aim to improve quality through proper handling and training. They encourage the participation of all employees in all quality improvement activities—thus making them totally committed!" The High Reliability program has two main elements. First and most important is the "Quality Control (QC) Patrol." QC Patrol monitors worker adherence to 5S, or "good housekeeping" rules,[7] and material handling procedures. The QC Patrol is made up of a group of "auditors"—all former operators—who police the plant and mete out disciplinary actions

---

7. The 5S system, common across the industry, is based on the Japanese model of good housekeeping that stresses orderliness, discipline and cleanliness. The five S's, translated from Japanese, are *Seiri* (Organization), *Seiton* (Tidiness), *Seketsu* (Cleanliness of self), *Seisou* (Cleanliness of the work station), and *Shitsuke* (Discipline). In practice, 5S demands that workers keep their workstations clean (many are forced to clean during their rest periods), are always dressed according to strict regulations, and follow behavioral rules in all common areas, such as the rest rooms and canteen.

for those caught violating standard operating procedures. Other auditors, particularly from HR, roam the plant with Polaroid cameras in search of violators, both in production and in areas around the canteen, restrooms, and employee lockers where workers tend to congregate and socialize. As described in the book's opening vignette, the QC Patrol maintains a large bulletin board outside the canteen where it posted the Polaroid pictures of workers "caught" in violation of dress or behavioral quality standards. One HR auditor admitted that workers go out of their way to avoid meeting her now. "We roam around the plant to talk to employees. If an employee smiles, then we talk to them. But since we've been identified as the ones giving out DAs [disciplinary actions], people try to avoid us."[8]

The second element in the High Reliability program is the Small Group, the firm's modest version of Quality Teams. The off-line teams make production-related suggestions for quality improvement. The groups are open to everyone, but in practice, the teams are made up almost entirely of engineers and technicians. Operators—90 percent of whom are women—seldom join these teams because most do not feel qualified to participate and are often intimidated by the engineers and technicians, who are almost all men. In addition, these teams usually meet off-line during breaks and many operators did not want to spend their few moments of rest in meetings with superiors. Engineers and technicians, on the other hand, are more comfortable with one another, often suggest projects for study, and have the freedom and autonomy from the assembly line to pursue such side projects.

PRODUCTIVITY   With engineers and technicians handling the bulk of production flexibility and quality programs, assembly line operators are subject primarily to the demands for continuously improving productivity. As one engineer noted, "The production engineer is concerned mainly with quantity; making target. They are also interested in quality, of course, but that's second." Production targets have been ratcheted up continuously since the plant came online. Output of HDDs grew from 355,000 units per month in 1996 to a targeted 1.5 million units per month in 2000 or a nearly fivefold increase in four years. To meet production targets Storage Ltd. operates around the clock, year-round. There are officially three eight-hour shifts, but with four hours of forced daily overtime there are, in practice, only two twelve-hour shifts. This is, in fact, a fairly common practice across the industry, although clearly in violation of the Philippine Labor Code (Article 83). It is also the major complaint about work at Storage Ltd.

---

8. This was quite apparent while walking through the plant with her. I noticed that passing workers would try not to make eye contact or would even abruptly turn and walk the other way when they saw us coming.

Belinda, a twenty-one-year-old production operator on the test line who has been working at Storage Ltd. for more than three years, complained:

> Things are run differently here. It's really strict. We have to work overtime, even if we don't want to, because we're the last process on the line. . . . If there's overtime, everyone works overtime. There's no one you can get to substitute for you because they're all tired too. . . . You have to sign an overtime slip. If you don't sign, or sign and leave, you're given a DA [disciplinary action] and you have to give a written explanation. If your reason [for not doing overtime] is not accepted, you're given a three-day suspension. . . .
>
> If we don't reach the target, they extend our hours even more, to double overtime ["OT-OT" from 8:00 A.M. until 10:00 P.M.] . . . and they increase the target every month. Everyday, our subleader says, "We're still negative [below target]," and if the subleader can't get us up to target, then the production engineer is brought in next.

Other operators also spoke of the forced daily overtime and the more-than-occasional fourteen-hour, double-overtime shifts. According to Irma, a production operator in the slider section, "The overtime is just too much. I only sleep five hours a day. I get home [from the night shift] at nine a.m., I'm asleep by ten. Then, I get up again at three or four p.m. just so I can do my laundry and clean up before I go in again."

Workers also mentioned other types of punishments for refusing to render overtime, such as not being allowed to work overtime for an entire week following the DA, even if one wants to. Others said they were not forced to work overtime, but, like most workers, complied voluntarily everyday. This hints at the need of many operators to work overtime in order to boost pay and make ends meet. This issue will be raised again in more detail in the section on wages and salaries.

During the twelve-hour shift, breaks are minimal. Workers are given an unpaid one-hour lunch break, two ten-minute "environmental breaks" (morning and afternoon), and one twenty-minute break during overtime. For those working in the Clean Rooms, it means they usually only leave the production area once a day for lunch. Famy, an encoder in the head and gimbal section, says that "even during OT, twenty minutes is not enough time to get out of the Clean Room to go the toilet. It's just too much trouble to get the bunny suit on and off. So we just hold it." Consequently, the most common job-related illnesses mentioned by operators are urinary tract infections and exhaustion. When asked why they ran only two shifts, the HR director said:

> Many American companies have three shifts, but three shifts means more people, more administrative support, longer canteen hours, and more busses. Right now, we spend four million pesos on buses alone. We have thirteen pick up points and seventy-five buses every morning and fifty-five buses every evening. Two shifts are just more cost effective.

The two-shift policy is also a tool for increasing internal labor flexibility. As mentioned above, Storage Ltd. hires only regular or permanent workers, who enjoy the legal rights of security of tenure and are thus difficult to fire. The use of forced overtime allows the firm to increase production without actually adjusting the total number of workers employed or decrease production without resorting to worker layoffs. The firm also minimizes its expenditures for per-head expenses, such as benefits packages, meal allowances, and bonuses.

In addition to daily forced overtime, workers normally work six days a week with a rolling day off. But when production demands, they are asked to come in seven days a week. During one stretch, the last quarter of 1998, large sections of the plant did not get a rest day at all for three straight months to meet the end of the year targets.

Conversely, workers were also subject to unplanned or unannounced forced vacation leave when, for example, parts in the slider section were unavailable for assembly production. If an operator has already used up her paid days of vacation leave, then she gets no compensation under the firm's "no work, no pay" policy. At an Employee Relations meeting (discussed further below), worker representatives repeatedly asked whether all workers could have Sundays off to spend time with their families. To this the HR director replied, "We need to work more hours and need to rotate, period. Every two months, you'll get a Sunday off. So just reschedule your dates. Go to church before or after your shift."

In a separate interview, I asked the HR director how the long hours and lack of rest days affect the workers. He commented, "Yes, some of the younger workers complain because they have to work on Sundays. They say that they have no opportunity to wash, to meet friends, to go to Mass. But once an operator works here for six months, they get used to it."

The Inside Game: "Disciplinary Management"
and In-Plant Commitment Strategies

Like others at the higher end of the industry, Storage Ltd. managers see building worker commitment as crucial to their competitiveness. And as noted in the introductory chapter, commitment is best understood through its three constitutive components: loyalty, attachment, and discretionary effort. However, because its flexible production processes are technology- and engineer-led, the firm does not rely heavily on enhancing loyalty among production operators, since their tacit knowledge and participation in problems solving is not crucial (Appelbaum et al. 2000). Rather, the focus within production is on boosting worker effort—secured primarily through technical control on the shopfloor and the disciplinary and surveillance practices of the HR department. The HR department is one of the largest departments,

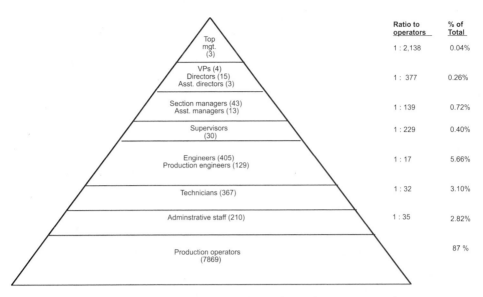

Figure 3.1 Staffing levels and workplace hierarchy at Storage Ltd.

with fifty staff members, and the only one completely run and staffed by Filipinos. Nearly the entire staff is made up of young, college-educated women.
The HR director, one of the few men in HR, is the highest-ranking Filipino
in the firm and the only non-Japanese in upper management (see figure 3.1).
The HR director is quite proud of this fact, but also admits that he has a
"shadow" Japanese counterpart, responsible for managing the forty-seven
Japanese nationals at the plant, and for keeping an eye on the Filipinos.

The HR department uses a range of tactics, which one HR staff member
aptly described as "disciplinary management." Disciplinary management relies, on the technical side, on data generated by the computerized assembly
line, which provides the "hard numbers" on productivity and quality performance per worker. On the social side, discipline is generally maintained
by strict enforcement of company rules. Every week 150–250 formal disciplinary actions (DAs) are meted out. A staff member commented, "Employees have a lot of disciplinary problems, especially in production, like sleeping
or neglect, which lead to damaged products. When they are under stress and
pressure from the production engineer or line leader, they sometimes lose
control." "Losing control" in this sense meant that operators would go
AWOL (absent without official leave), answer back, and neglect their work.

HR reports listed twenty-one "common offenses committed," including
going AWOL, failure to follow work procedures, improper wearing of uni-

forms, wearing makeup, clocking in/out for others, and sleeping on the job. Sleeping was very common, particularly at the high magnification microscope sections. Because the stress and eye strain was so intense (and because they could get away with it), many operators "took naps" at their stations with their eyes and faces still pressed to the microscopes. Sleeping has become such a problem that the penalty had to be increased from a one week suspension to a one month suspension, or even immediate dismissal if the offense constitutes, "a deliberate form of sleeping with malicious intent."[9]

SALARY, BENEFITS, AND PROMOTIONS   Recognizing the high demands of the job and to help stave off complaints and possible worker actions, the firm does provide some positive incentives. First and foremost, take-home pay is relatively high. While operators' base pay is only minimum wage (P200/day or about five dollars), the company does pay for overtime (125 percent of the base pay rate), which means workers' monthly take-home pay, about P6,000 to P8,000, is more than most local factory workers. Nonetheless, workers still said they could only manage to make ends meet with the overtime pay. Famy, a single operator living in a boarding house, noted: "I make P211 a day and it's not enough. I spend P1,200 a month on food alone. I can't cook, so I'm forced to buy my meals [from vendors]. Sometimes, when I don't have any money left, I just don't eat."

Everyone receives a Christmas bonus (equal to one month's base pay) and a bonus based on overall company performance (usually equal to one month's pay). On an individual level, there is a perfect attendance bonus for no tardiness or absences for three months. There is a free shuttle bus, free uniforms, a subsidized meal allowance, a canteen, a medical clinic, and twelve days of sick and vacation leave a year. As one staff member put it, "They are provided everything when they begin work here, what else could they ask for?" In fact, Storage Ltd. does provide all the legally mandated benefits, which exceed those offered in other firms and industries.

Conspicuously absent, however, are any seniority-based pay increases or meaningful promotional avenues for operators. Starting salary is minimum wage. Belinda, quoted above, is typical: she has worked at Storage Ltd. for more than three years, putting in twelve to fourteen hours a day, often seven days a week. Yet, her base pay remains P215/day or only P15 above the daily minimum wage. At an Employee Relations meeting, one worker suggested a "loyalty" raise or promotion after three years of service to the company. To this the HR director replied, "We're labor intensive. We must focus on pro-

---

9. When asked to explain the term, the HR director replied that it involved workers who would actively find hiding places to sleep, while arranging with other workers to cover for them and act as lookouts. The disciplinary policy demonstrates the lengths that management will go to punish *collective* and *deliberate* resistance.

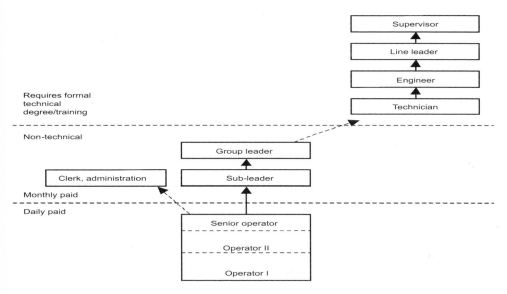

Figure 3.2   Career ladders at Storage Ltd.

ductivity and we can't promote everyone. 'Pay for performance'—that's our policy."

There is also very little upward mobility for operators (see figure 3.2). The three grades, Operator I, II, and Senior Operator, are based on yearly performance appraisals, but have little difference in pay. Practically, there are only two levels above operator to which production workers can aspire: subleader and group leader. Operators must show exemplary performance for two years to be considered for a position as a subleader, a monthly paid position that earns only P10 per day above that of operator.

In an Employee Relations meeting, production engineers and line leaders, who act as de facto supervisors, complained that the company policy to promote no more than 5 percent of operators was too strict. They argued that they had more than 5 percent of operators obtaining an excellent rating according to the company's own Bell Curve formula based on yearly performance appraisals. To this, the HR director replied, "Last year, we had too many promotions. This year, we need to rationalize promotions. That's the way the cookie crumbles."

When operators were asked in interviews how far they could rise in the company, all replied that group leader was the highest level they could reach. Only the technician interviewed said she could go further, all the way to supervisor. While operators could theoretically become technicians, such mobility is effectively made impossible by the firm's education requirements. For reasons discussed in the next chapter, Storage Ltd. does not hire operators with more than two years of college. Yet technicians and engineers are required to have technical degrees, which usually take three to four years to complete. An operator who wants to be technician must first complete training, then maintain excellent performance doing the job of a technician for six months to one year before being formally promoted to technician. In addition, the current production schedule does not allow workers to either take time off the line for such training or provide enough rest days to attend a course outside of working hours.

EMPLOYEE RELATIONS   To ameliorate such demanding working conditions and limited mobility, the company endows the Employee Relations (ER) section with the daunting task of promoting loyalty, attachment and discipline, which it does through a range of practices. First, they conduct companywide "unwinding and morale-boosting activities" throughout the year, including a Christmas Party, Dance Contest, Summer Outing, and a Sportsfest. They also encourage employees to report to them any misconduct and at least ten reports of mistreatment by superiors are called in anonymously each week. As one ER staff commented: "Not all employees are courageous enough to approach ER staff. So we try to establish personal contacts within different parts of the plant. To get information. . . . you have to establish relationships with key people. This way, we can relay the message that ER staff are not monsters."

Probably most important, the ER section convenes a monthly ER meeting, in which worker representatives meet with the HR director and staff. This is the workers' sole forum for bringing up issues and complaints, and they do come with long lists. Ostensibly, the ER meeting is set up "to provide an atmosphere of open communication where employees discuss issues regularly with peers, subordinates, and superiors." The structure and activities of the ER meetings mirror those of a labor management committee (LMC). But as the HR director noted, "We don't want to call it an 'LMC' because it is a term associated with unionism. We want to temper the rhetoric and use a more subdued, discrete term."

The meetings are split into three separate sessions with representatives from each section in the plant: one for operators, one for sub- and group leaders, and one for engineers and staff. Despite their overwhelming numbers, operators get only thirty representatives, about the same as the other two groups. Representatives are selected by their peers and asked to hold section-level meetings to collect comments. The representatives are also asked to report to HR any problems or violations they witness and act as examples for other employees.

At the meetings, many complain about the quality of canteen food, dirtiness of the toilets, and the rudeness of shuttle bus drivers. Questions often come up about adding new shuttle bus routes, getting new uniforms more often, and the precise definition of "sleeping on the job."[10] A few ask more pointed questions about the lack of rest days, the lack of seniority pay, and the difficulty of promotions. Yet employees raise few questions about pay and none about the heavy overtime or production itself. In this regard, the general level of fear and self-discipline among workers keeps many big issues off the "open" agenda.

Although workers interviewed mentioned the ER meetings as their one forum to speak up, there are few illusions about what is appropriate to discuss and who is in charge. Kiko, a warehouse operator, is also an ER meeting representative:

> All those that want to communicate with management have to go through us [the committee]. We [the workers] are all members and the HR director is the head. He's also the one who gives the orders [*nagdidikta*]. The only thing we can do is air gripes. We're like children that have to ask permission. If they say yes, then it's OK. But if they say no, then no.

While falling short of a genuine forum for worker "voice," the ER meetings do serve an important, less explicit function. As the HR director put it: "Employees must feel like they can speak up and must know what's going on. We want to avoid a situation where the employees organize a union. We don't need a third party to communicate with our employees. And I am happy to say we are union-free."

The antiunion strategy of the HR director was clearly shared by the entire department. As one ER staff member admitted, "[The HR director] told us, 'If a union is formed here, you [the ER staff] will all be asked to resign' since having a union would mean that we had not done our jobs."

Finally, the ER staff conducts exit interviews with all employees who quit voluntarily or are "asked to resign." When asked about what "asked to resign means" an ER staff person responded:

> Here at Storage Ltd., we don't "fire" employees, we encourage them to resign. We say, "it would be better if you resigned." We then ask them to write their own resignation letter. We don't use a form letter, because it could be used as a basis for an employee to claim illegal firing. We just present them their records as a basis for encouraging them to resign."

---

10. Reporting and what constituted sleeping on the job seemed to be a big issue, since it was common among technicians and engineers during the night shift, and operators were afraid to report them. Operators asked if they could have access to a digital or Polaroid camera because if they reported a superior, they had to have a photo for "really good proof."

The panoptic work regime, then, with its "flexible Taylorism" and disciplinary management, provides little room for workers to slow down, let alone participate in, work organization. Thus, the firm does not need to foster worker loyalty or offer positive incentives to boost worker effort. As Delbridge (1998) notes, under such a tightly efficient system, even a "work to rule" job action by workers would have little effect on output, since production efficiency has nearly been maximized by the process engineers and built into "the rules." Because the work pace is technically controlled, worker effort is reduced to a willingness to work, rather than a willingness to work hard. Yet the firm still requires a stable and committed workforce, which it secures primarily through its worker control and attachment strategies. Earlier, I discussed the firm's sophisticated in-plant approaches. But, in fact, to foster worker attachment to job and company, Storage Ltd. relies heavily on external strategies of localization, recruitment and surveillance that are integral to the panoptic work regime and the generation of alienative or "coerced commitment" which will be taken up in detail in the next chapter.

Finally, the case of Storage Ltd. also highlights the growing gap—in pay, autonomy, and occupational mobility—between women production workers and male technicians and engineers. This gendering of high technology and worker stratification in the face of technological upgrading is also evident in the other two high-tech case studies and will be discussed at length below.

Integrated Production and the Peripheral Human Resource Work Regime

> On a large electronic board hanging prominently in the canteen, the stock quote direct from the New York Stock Exchange flickers continuously by. Operators, dressed head to toe in full bunny suits, pass briskly, aware of the company's "two-steps-per-second" policy. Passing through the air shower, they enter the nearly dust-free production area. On the floor, machines, rather than workers, dominate. Their low hum fills the room. In the die attach section, minute silicon chips are attached to a lead frame while operators tend seven identical machines and have little contact with one another. Rather than furiously assembling components by hand, workers spend most of their time monitoring, and being monitored by, the automated machines. The enormous machines, each piece wired to its own computer, assemble the chips while feeding information into the controller. Workers glance casually at the large screens instead of squinting into microscopes, making minor adjustments to the product feeder to prevent jamming.

At the end of the line, process engineers gather at several computer terminals, tracking the output and ensuring the product remains within the quality bands. Above their heads a large output board displays the same information for those on the other side of the glass walls. On the board, just below overall targets and actual production, each operator is listed by name, followed by her output, number of disciplinary actions, and number and type of defects. Supervisors update the board every two hours. Alongside, a large quality awareness board reminds workers: "Misorientation is mostly a human dependent defect."

field notes: 16 Sept. 1999—Integrated Production, Cavite

Integrated Production represents a third organizational pattern, the peripheral HR work regime. At its core, the system relies on engineer-led production flexibility but unlike the panoptic regime, it combines its technology strategy with a strategic HR approach that stresses positive incentives (Legge 1995). Patterned after its corporate parent, this work regime closely resembles the Silicon Valley variant of the American HR model that promotes individual loyalty and ritual participation, rather than simple attachment (Appelbaum and Batt 1994; Katz and Darbishire 2000; Milkman 1991). But as the case study will show, the peripheral HR model falls far short of fulfilling the promise of worker-centered high performance. And in exposing the downsizing and persistent gender segmentation associated with technological upgrading, the case study also demonstrates both the threat and unrealized potential of automated high-tech production for women production workers (Kelkar and Nathan 2002; Wright 2001).

Integrated Production is a wholly owned subsidiary of an American firm assembling advanced integrated circuits, or IC chips. To meet the cost, quality, delivery, and reliability demands of the IC market, the firm focuses on automation. The firm's $200 million state-of-the-art facility opened in late 1996 and management boasted of having the industry's latest machines, each costing upwards of $1.5 million. As at Storage Ltd., all production takes place in Clean Rooms. But unlike most other assembly operations in the Philippines, it has an on-site chip design facility to reduce cycle time and allow for customization. The plant is located on over four hectares in one of the premier, private special economic zones (SEZ) in the province of Cavite.

The firm maintains a workforce of 1,373 permanent employees, of whom only 55 percent are production workers. In one month, Integrated Production exported about 13 million chips valued at $35,422,028. While the average monthly salary of a production operator—P6,850 or $180—was similar to one from Storage Ltd., output per employee was $25,799, or more than three times that of Storage Ltd. or Discrete Manufacturing.

Flexibility, Quality, and Productivity Strategies

FLEXIBILITY   Integrated Production builds as much flexibility into its products and processes as possible so that their customers can make changes even late in the product development cycle. The firm maintains dedicated and flexible production lines for different clients, lines which can be quickly rearranged and are electronically monitored in real time by the home corporation as well as by customers. As at both Storage Ltd. and Discrete Manufacturing, flexibility is engineer-driven, achieved primarily through the use of programmable machinery and sophisticated manufacturing information systems. The firm employs more than three hundred engineers and technicians—approximately 20 percent of the entire workforce—to keep production running.

Operators working the expensive machines must at least understand the basic production process and go through extensive initial training and certification. According to the training manager, operators do not reach their peak productivity until after six months. Workers are trained to operate one "mother station," then are cross-trained on two other processes. While the company had initially embraced more extensive multiskilling, it had begun limiting operators to a maximum of three certifications. According to the training manager this was because the processes had become increasingly complicated and "there is too much to forget." In fact, operators were no longer moved around extensively, spending nearly all of their time at their "mother station." Maria, who has been with the firm for one year, noted:

> I can operate almost all the machines. We have seven machines and I'm certified on almost all of them. . . . But right now, there is one machine that no one else but me operates. And wherever that machine goes, I have to follow it. Even if it goes to the PM [production manager] to be recalibrated, I have to follow it. It's like having a husband.

While Integrated Production's assembly lines are already quite automated, a new project, the A-line, was being developed for a new level of automation that would entirely eliminate operator handling of individual chips. Although the A-line promises to put the Philippine plant at the industry's cutting edge, the elimination of operator handling will likely lead to worker layoffs. As one HR staff noted, "The [A-line] project really means full automation. It will begin in the next two quarters and our operators are afraid."

For remaining workers, however, automation and integration also demand increasing technical knowledge and capabilities. With the introduction of the A-line, Integrated Production, like Discrete Manufacturing, has created a new shop floor position: the technical operator. Already, Integrated Production has a relatively low percentage of shop floor operators: only 55 percent of total employment (versus 87 percent at Storage Ltd.). The new technical operators will still handle basic production, but also take on many of the responsibilities of technicians, performing all minor maintenance, repairs, and adjustments. These technical operators are targeted for promotions to full technicians and possibly, engineers.

The development of the technical operator position has the *potential* for building in more vertical mobility for women operators by fusing the job and promotion ladders of operators and technicians that are, in many companies, kept separate.[11] If women operators were to gain access to these jobs, their increased skills and mobility could provide more decision-making power and help break down gendered notions associated with technology and technical positions that have helped keep women out of technical work (Ng and Mohamad 1997).

Unfortunately, such potential remains unrealized. First, the firm requires these new technical operators to have at least two years' training from an accredited electronics training school. In fact, managers, operating under gendered assumptions about women's technical (in)capacities, do not fill these new positions by training current operators, nearly all of whom are women. Rather, the firm recruits directly from engineering or technical schools, where almost all the trainees, and thus all the new technical operators, are men. The male bias in some sense reflects a technology bias and the wider gender segregation in Philippine education that funnels a disproportionate number of men into technical education.

Indeed, the persistent and strict gender division of labor makes upward mobility for current women operators extremely difficult. Overall, the firm has one of the most balanced gender ratios of the companies visited: 38 percent male and 63 percent female. But when broken down, the gender segregation becomes clearer. All nine directors and 90 percent of the engineers and technicians are men. Conversely, 85 percent of production workers are women.

In other ways, technological upgrading may already be shifting management's gendered notions about electronics assembly, effectively "masculinizing" the work and the workforce. The HR manager explained that, in the early days, women dominated semiconductor assembly in the Philippines because "everyone followed the dictum that women work better with small things." But with the increase in the number of technicians and technical operators, "guys are beginning to gain ground." He explained that now, the new automated machines are not as "gender specific" and that production has become "less dependent on the fine skills of women. . . . The machines have large screens instead of microscopes so [operators] don't need 20/20 vision or the patience to sit at a scope." To account for shop floor changes, he

---

11. Integrated Production has five classes of operator, with the top of the job ladder ending at trainer. Technicians, on the other hand, can rise to the engineer and even management levels.

also noted that now men, with fewer options for work, have to learn "women's jobs." He went on to explain why women do not make ideal workers: "Women get sick, they need maternity leave, have monthly periods." Men, on the other hand, were seen as "stronger" and able to work longer shifts. The rationale for hiring more males will be taken up again in the discussion of the compressed workweek below.

QUALITY   Raising product quality has been another driving force behind automation. As at Storage Ltd., the core of Integrated Production's quality program is a manufacturing information system that tracks production using statistical process control (SPC). Rather than active participation in problem solving, workers are expected to simply check the SPC readouts and call over an engineer if there is trouble. Workers are also subject to daily quality audits. Maria explained the system this way:

> We're always audited. . . . They check housekeeping [5S], the place has to be kept clean because the things we produce are very sensitive. We have to follow the specifications; they're like our bible. Whatever is written in the specs, you need to follow. If you don't follow, you get an ITR [inspection trouble report].
> For example, if you fail to write something down and you're audited, then you get an ITR, and so does your supervisor. They call a meeting, so you don't repeat it. If you repeat it, you get an ITR and a discrepancy. It's double. They issue all the ITRs with the supers, so it's embarrassing for the supers, and they [the supers] get really strict. I don't mind the ITR, but they shouldn't involve the super. It's embarrassing.
> They audit all machines, twice a day, once during the day shift and once at night. . . . When an audit is about to begin, we yell to each other, "Audit! Audit's coming!" and so we arrange everything. So it's tense and stressful. I get so flushed explaining my performance, even if I've done nothing wrong.

PRODUCTIVITY   The firm's productivity also relies on the manufacturing information system and a total productivity maintenance (TPM) program aimed at keeping the machines running full tilt. The TPM program centers on TPM teams that solve specific quality or productivity problems or implement particular improvement goals. Theoretically, these TPM teams include operators as equal team members. But, practically, the growing technical sophistication of the machines has meant that operators play only a minor role. One operator discussed her involvement with the TPM teams: "Technicians, process engineers, supers [supervisors] of technicians, and managers of technicians [attend meetings]. . . . First they meet, then they tell us what they talked about, then they ask us what we think. So it's only at their next meeting that we are able to speak." Another operator also commented: "Management only consults with supers and engineers. Sometimes we have meetings and they just inform us, but we tell them if we can't do it."

Rather than deep worker involvement, workers experience the pressure for high productivity mainly through work organization and performance targets. A core organizational strategy is the firm's compressed workweek, a policy patterned after their corporate headquarters.[12] Under this policy, workers work twelve-hour shifts for four days, then have three days off. Integrated Production is the only electronics firm with official government clearance to operate a compressed workweek. Nonetheless, many firms, like Storage Ltd., use de-facto twelve-hour shifts through daily forced overtime.

To win workers over, the firm presented the compressed workweek as a "family friendly" policy. According to the ER director, "The compressed work week jibes well with Filipino society because the Philippines is a family-oriented culture" and employees can enjoy more family time. Management also referred to the twelve-hour day as a reason for the shifting gender balance at the factory toward more male workers. According to the HR manager, "With the long shifts, we need more stamina and not agility. . . . Guys are strong and can work for twenty-four hours. With the right pay, they will do anything."

Although quite flexible, Integrated Production is a high-volume producer. The firm uses classic, Taylorist methods such as time-and-motion studies together with more advanced SPC to set productivity targets. Agnes, a test operator, explained:

Each operator doesn't have a target. It's a group target, every quarter. We do have daily targets: 100,000 units a day for two shifts. But if you don't reach it, they don't do anything. But there is a quarterly quota. This quarter we need to produce seven million units, just [our section], and if we don't finish, we have to work on Christmas. No shut down.

It's hard work. You get so many sermons. Because sometimes, FVI [final visual inspection] is swamped with so many bundles, then it's overflowing. Then you have to work fast. But if they find rejects, then they give you an ITR.

At times, the work pace can be quite pressured. Another worker explained,

Normally, I work only one machine. Sometimes I have to handle two or three machines, but you have to do it. Once, it was OT and I was confused and had to handle three machines. Loading one, then I run to the other lot to take it out, then run to the other to start its lot. It's hard; I almost don't get to eat.

However, most operators said that reaching the daily target is not too difficult, particularly if there is no machine downtime. Clara noted: "It's a

12. Integrated Production, to maintain high quality, had long resisted moving assembly offshore. An HR staff member confided that when assembly was still conducted in California, the long hours, fast pace, and high demands on workers earned it the nickname, "the sweatshop of Silicon Valley."

happy place and not too much pressure. Because your job is just to watch the machine that's testing units used in computers. You only have to test if they pass or not so you do nothing but watch the machines."

Strategic Human Resource Management (HRM) and Antiunionism

Although Integrated Production clearly focuses on "hard" TQM tools in production, they do not neglect the "soft" or social aspects of work. The firm's HRM strategy draws on the parent corporation's employee relations model, complementing the rigors of production with positive incentives. This strategy aims to secure employee commitment and prevent worker dissatisfaction, which management views as the prime motive for union organization. In this respect, Integrated Production is quite similar to the large, nonunion American firms that have adopted HRM to substitute for unionism (Foulkes 1980).

As described by Foulkes and updated by Ruth Milkman (1991) in her study of Japanese firms in California, the American nonunion model of industrial relations (IR) is an intermediate case, in between traditional Fordist mass production and more full-blown, Japanese lean production. Under the Fordist model, decision-making is highly centralized, work rigid and deskilled, pay is based on jobs and performance, status distinctions between workers and management are clear, labor relations are adversarial, and unions are strong. At the other end of the spectrum, under the "idealized" Japanese lean production system, worker participation, communication, and job security are high, there is extensive job rotation, pay is seniority based, status differentials are muted, labor relations are cooperative, and unions, while present, are weak.

In contrast, Milkman differentiates the American nonunion model as offering more worker participation and communication than the Fordist model, but far less substantive involvement than Japanese lean production. The American nonunion model also sets itself apart in the prominent role that HRM policies play in the overall work regime. Here, a well-developed communication program, an internal labor market, above-average pay and benefits, and job-security distinguish the model. Crucially, Milkman (1991, 72) adds, "One key feature that differentiates this model from both Fordist mass production and the Japanese model is the absence of any form of unionism and a strong managerial commitment to avoid unionism."

Integrated Production takes a similar approach, focusing on improving the company's bottom line by boosting the full spectrum of worker commitment despite the lack of substantive participation, lower job security, and a conscious effort to thwart unionization. Firm representatives spoke quite explicitly about their antiunion attitudes and the firm's union substitution policies. The HR manager stated flatly:

A semicon company's worst nightmare is a union . . . and it is the primary task of HR is to insure there will be no union. . . . We give extraordinarily high pay and benefits because we want people to be flexible. We want to be able to move people around and change their schedules to respond to market demand. Unions tie up management and don't let us be flexible.

Yet, in fact, Integrated Production did not pay extraordinarily high wages. As noted above, the average salary for production workers was only P6,850 per month, or far less than Storage Ltd. (which had the same base wage but much higher take-home pay due to forced daily overtime), and nearly half that for Discrete Manufacturing. Rather, Integrated Production uses a comprehensive HRM strategy trumpeting "positive discipline," coupled with external labor recruiting and control policies detailed in chapter 4.

In terms of internal strategies, the firm tries to maintain a positive corporate culture and atmosphere of trust by creating a "good work environment." For the ER manager, this means that the firm is "really like an American company [in America]—there's really toilet paper in the bathroom, really free coffee all the time." Creating a good atmosphere also includes trying to decrease the sense of hierarchy and status differentials. For example, one operator explained,

> Even with people higher than us, the supers, I can joke with them now. It used to be "sir, sir," now it's "hey Bob, hey someone." . . . It's like we are a clique (*barkada*), all friends. We can joke with them, like, "How are you love, how are you sweetheart." Even the Americans, they don't want to be addressed as sir; it's Willy or Sam.

Behavioral training also plays a central role. In lieu of substantive participation, production workers go through "total quality commitment" training designed to build a "quality culture" at the plant. The trainings focus on promoting and personalizing "quality values" among workers. The head of the training staff noted, "each worker must sign a Personal Commitment to Quality that states that the worker will 'live by quality values' and added, "this is meant to dramatize and ritualize their commitment to quality." She also noted proudly that rival companies, particularly Japanese ones, do not invest in or conduct nearly as much behavioral training as Integrated Production.

Direct supervision is also not constant and Integrated Production maintains a relatively low supervisor-to-operator ratio of one to fifty. Compare this to Power Tech where the ratio was one to seven and at Storage Ltd. where the ratio of workers to engineers (who act as de-facto supervisors) was one to fourteen. One manager explained that while they use pure gold wire in production, there were explicitly no cameras in the Clean Rooms because "we trust employees. They go through behavioral training and we teach them honesty." However, the same manager then added, "but we also have large glass windows all around the production area" to keep an eye on them.

PAY, BENEFITS, AND MOBILITY   On balance, Integrated Productions does try to provide primarily positive incentives to engender worker commitment. Yet at the same time, the firm's cost-conscious HRM strategy focuses squarely on individualizing the employment relationship and making salaries and incentives performance- rather than seniority-based. For example, pay is calculated at a monthly salary base rate rather than at the more typical daily rate paid by most other firms. Being a monthly-paid "salaried employee" is considered a notch up in status from a daily-paid laborer. Yet the starting monthly rate is P5,700 per month, hardly more than the take home-pay of a daily-paid minimum wage earner.

Added to this base rate are various supplements. First, the firm pays a legally mandated thirteenth-month bonus as well as a voluntary extra fourteenth-month pay, given near Christmas. All employees also receive at least fifty shares of company stock when hired and additional shares as a yearly bonus. Next, workers receive a quarterly production bonus based on new product revenue, and, finally, individual bonuses based on performance, attendance, length of tenure, and "attitude."

In terms of benefits, all employees receive fifteen days of vacation and sick leave a year, medical insurance, medicine, daily meal, and rice subsidies; free shuttle service; stock purchasing options; and an education reimbursement plan. But the benefit most boasted about was the company's computer purchase program. After one year as a permanent employee, each worker is entitled to a US$1,200 reimbursement toward a personal computer. Considering that the reimbursement amount equals over 75 percent of a Filipino operator's base yearly salary, the benefit is indeed something to brag about. But the policy is also cleverly designed to promote job attachment. Employees do not own the computer outright for an additional year, and if a worker leaves before completing a second full year, she must return it. In the minds of many employees, the free computer puts the firm above nearly all other companies, including ones that paid substantially higher salaries.

Another attractive aspect of the job for many workers is the potential for internal promotions; many operators hoped to take advantage of either technical and nontechnical internal job ladders. On the technical side, if operators have the proper educational background, they can take advantage of internal training programs. They can also tap into the firm's educational subsidy and return to school for a technical degree. However, as noted above, to fill the new technical operator positions, the firm has not hired from within, but has instead recruited directly from technical colleges.

More plausible for most operators is a horizontal move to administration or a nontechnical job ladder to a supervisory job. Here, operators can move to office work as administrative assistants or be trained internally to become trainers, Quality inspectors, or line supervisors. All operators interviewed felt that they could rise in the company. In fact, because a majority of operators have such high education levels but face a very slack external labor market, many took jobs as factory workers in the hopes that they could eventually move up in the company or, at least, out of production.

EMPLOYEE RELATIONS   Within the HR department, the ER staff coordinates a number of programs to provide a "positive atmosphere" and handle traditional labor relations. Their mandate, as explained by the ER manager, is clear: "The goal of ER is to provide a worry-free shop floor for supervisors."

First, the company provides social activities, such as sports programs, an annual Christmas party, a Family day, and activities around Lent. Such programs significantly impressed employees. For example, one worker stated,

> We have lots of company outings. We have pizza party every quarter. We also have an Integrated Production band. Everyone is invited to join, as long as they have a good voice. There are two bands in Integrated Production. They performed at the last pizza party. The nice thing is, it's not just like a company, but a family.

The firm also has a comprehensive program for management-worker communication. There is a quarterly communication meeting where the general manager (GM) presents policy changes, overall performance numbers, and the amount of the quarterly bonus. There is also a weekly *ugnayan*, or connection, meeting between the GM and eight to twelve employees selected by the ER department. At the departmental level, there are weekly meetings to discuss production issues and general working conditions. But like the total productivity maintenance (TPM) teams, these meetings are usually dominated by managers, supervisors, and technical staff, and mainly used to inform operators of changes.

Outside of production, the ER has set up a "Talk" program with suggestion boxes, a monthly newsletter, and short section meetings before each shift. The program has helped create a positive company impression. One operator, when asked what she liked about her job, said, "What's good is at Integrated Production, we have 'Talk' where you can voice your opinion. All you have to do is write a letter. . . . There are suggestions boxes everywhere."

Finally, the ER manages an employee appeals process to allow employees "a course of action to help resolve problems in the workplace." The employee appeals process follows all the procedures of grievance machinery, but as the ER manager explained, "We don't call it 'grievance machinery' because it sounds like a union."

In fact, the company saw the ER clearly as a substitute, not only for a union but also for collective worker representation in general. Indeed, there are no meetings in which workers choose their own representatives. Management has been quite careful to allow for a variety of channels for individual communication without any sort of collective representation, even in a typically management-dominated forum such as a labor management committee.

By most accounts, the ER staff has been quite successful. When asked about the types of employee grievances that are raised, the ER manager explained that most are "miniscule . . . and that's how we like it." These "miniscule" problems usually include complaints about the canteen food, the shuttle buses, and small problems with employee paychecks. He then added, "They [operators] have no major complaints because we have a very proactive strategy. Employees don't feel they are exploited. We maintain issues at a trivial level." With their broad range of tactics, the ER staff has deftly created a communication program that is viewed positively by workers and also serves to nip problems in the bud through extensive information gathering.

LABOR RELATIONS AND ANTIUNIONISM   Integrated Production also takes a very aggressive stance toward heading off any collective action by workers. In their official "Philosophy on Employee Organization," the company is quite straightforward: "We prefer to deal directly with our employees as a team in an environment of cooperation and trust without the animosity that can be caused by the presence of a union. . . . We want to avoid the possibilities of work stoppages that could lead our customers to look elsewhere for their needs."

The company also includes "union activity" on its list of employee violations that are subject to disciplinary action, despite the clear illegality of this company policy.

A centerpiece of Integrated Production's strategy is a labor relations training program for its supervisors. The three-day training, even though it covers useful elements for worker protection, such as laying out relevant labor laws and worker's rights, is explicitly designed to keep unions out without crossing the line into illegal acts that might subject the firm to legal action. A major portion of the training is to help supervisors recognize "the signal signs for unionism." The Labor Relations Training Handbook includes a section titled, "Determinants of Unionism," which details, in a thirty-page overview, the "warning signs" of union activity. These include "the grapevine suddenly goes dead; groups of employees suddenly become quiet when you approach; and employees begin using new technical terms such as protected activity, showing of interest, demand for recognition, and unfair labor practices."

The handbook also details, "organizing tactics utilized by unions" such as "use of an 'undercover agent' or 'salting'; home visits; internet web sites; e-mail spams; and picnics, barbecues, lunches, dinners, rallies, or similar gatherings" ([Integrated Production] Human Resources n.d., 53).[13]

The training goes on to cover a long list of suggested interventions and "views and opinions you may communicate to employees." These included the distribution of "copies of articles and newspaper clippings reporting such matters as union related violence." Finally, referring to the Philippine Labor Code, the training carefully advises supervisors on what they can and cannot legally say or do regarding job security, outside organizers, strikes and picketing, and the circulation of union petitions or membership forms.

While the firm's official statements and documents are very careful to avoid overstepping any legal bounds, the message is quite clear. When asked what the firm's view of unions was, one operator said very matter-of-factly, "Oh, there's no union here, it's forbidden. They say that right at the orientation." Integrated Production's active and sophisticated antiunionism strategies, in combination with external strategies discussed in the next chapter, clearly have negative implications for improving workers' collective power.

Purchasing Commitment

As in the case of Storage Ltd., it is clear that rank-and-file workers do not play a substantive role in decision-making on the shop floor, although production does require a stable and committed workforce. However, the firm builds worker loyalty, effort and attachment using a strategic HR approach that emphasizes positive incentives, while individualizing employment relations, thwarting collective organization, and leaving undisturbed a hierarchical management structure.

First, management is not interested in worker "empowerment" and does not promise, like Discrete Manufacturing, to "throw away old concepts about power and authority" in the pursuit of high performance. Rather, it paints itself as a "quintessential entrepreneurial company," which, according to their corporate profile, " . . . encourages individuals to do what it takes to get the job done, provides them with the proper tools to achieve these objectives, and rewards them for their efforts. These individuals have made, and continue to make, [Integrated Production] successful."

The peripheral HR work regime is thus designed to address individual

13. Interestingly, many of these tactics, such as the use of e-mail spams and internet sites, while popular in the United States, are not yet practiced in the Philippines, suggesting that some of the training materials may originate from the corporate head office or other U.S.-based labor consultants.

worker issues and provide, at most, individual voice to head off potential problems. Management has also been effective in promoting individual loyalty in part because it creates a *sense* of worker control and employee satisfaction—although not actual decision-making power—through its assortment of social functions and morale-boosting activities, like letting workers use the restrooms without prior permission. Management's sophisticated communications system also allows workers to feel that they are listened to without providing workers collective representation or fundamentally challenging management's control over work.

The firm's strategies uses to promote individual loyalty, in conjunction with "hard" TQM and electronic monitoring tools in production, have been quite effective in also securing worker effort. The fully automated production reduces the scope of workers' ability to shirk and sets a floor on their effort bargains. However, management recognizes that strict technical control alone may lead to dissatisfaction and turnover. Managers are also keen to avoid high levels of dissatisfaction to prevent any potential basis for union organizing. The firm thus uses a pay-for-performance scheme and again turns to behavioral training to help align the firm's and the workers' interest in high productivity.

Finally, the firm is clearly interested in fostering worker attachment in order to minimize the costs associated with turnover and absenteeism. Its strategy, unlike that of Storage Ltd., is to keep workers satisfied at work and thus more attached to their jobs. Yet as discussed in chapter 4, the firm chooses to hire somewhat "older" workers with higher levels of education. And despite the positive incentives, Integrated Production's workers are not necessarily more satisfied or attached to their jobs than those at Storage Ltd. To more fully understand workers' satisfaction and attachment, we must explore their backgrounds, past work experiences, options in the local labor market, and the effects of their work on their lives outside the plant. These issues and how they affect worker attachment are taken up in detail in the next two chapters.

In sum, Integrated Production has been relatively successful in eliciting worker commitment by fostering a "new collectivism," one that scrupulously avoids the creation of a collective worker identity, but rather one that secures "*individual* employee commitment to *collective* organizational goals" (McLoughlin 1996, 305 emphasis added). Ironically, the most apt description of the firm's approach comes from a section of its own training manual called "Selling Leadership Style": "Most of the direction is still provided by the leader. He or she also attempts through two-way communication and socio-emotional support to get the followers psychologically to buy into decisions that have to be made" ([Integrated Production] Human Resources n.d., 9). From its own materials, Integrated Production itself provides perhaps the textbook definition of "purchased commitment."

Discrete Manufacturing and the Collectively Negotiated Work Regime

Along Line Two, workers are busy assembling plastic power transistors for use in everything from cell phones to home appliances. The preponderance of older, mainly female operators move about freely, loading machines, monitoring their stations, and chatting with other workers while they collectively churn out some eighty thousand chips per shift. The factory floor is open and the production lines progress from the humming of automated bonding machines to the heat and grime of the semiautomated plastic molding section to the pounding of the loud, manual flashing and cropping stations. A bulletin board filled with fishbone charts and Pareto diagrams trumpets the success of the section's new quality team program; it is ignored.

The operators resemble hospital workers; they wear protective light blue head coverings and lab coats, but also wear makeup and jeans. Most have their facemasks pulled down around their necks to easily breathe and talk. Because the chips are lower-tech and not so environmentally sensitive, workers do not work in Clean Rooms, and they do not have to wear full-body "bunny suits," nor do they have to "plug in" to their workstations to discharge static electricity.

As I walk the floor with Ramon, the operator turned supervisor, we stop and talk with workers, who tease him because he is young and single. Elsie, the line leader who has been at Discrete Manufacturing for seventeen years, jokes that she can work all fifteen machines in her section with her eyes closed. In the cropping section is Jose, the twenty-nine-year-old union shop steward and one of the few men in the area. Jose assures me, giving Ramon a pat on the back, that Ramon is "one of the good ones." Ramon tells me that keeping a good working relationship with the stewards and the veterans is the key to keeping things running smoothly and maintaining a "good culture" in his sections. But he does lament that he's having disciplinary problems and lots of absentees. When I'm ready to leave, some thirty minutes before the eight-hour shift ends, a number of workers are already gathering near the exit doors, ready to "swipe out" for the night.

field notes, 29 November 1999,
Discrete Manufacturing, Laguna

Discrete Manufacturing, which represents the collectively negotiated work regime, differs markedly from the other cases of high-tech flexibility. It is one of the few unionized multinationals in the industry, disproving claims by many in the industry that unionism and electronics are simply incompatible. In part because of the union's bargaining power, the firm's negotiated regime pursues flexible production using a three-pronged strategy: new production technology, work re-organization, and the promotion of active worker commitment. But management also wrestles with labor control and thus also uses the introduction of flexible production as a disciplinary tool to elicit greater worker effort and divide union strength. Still, workers and union officials participate, trying to negotiate labor process change, capture a measure of worker control, and defend shifting positions of authority.

Discrete Manufacturing is a wholly owned branch plant of a leading European electronics multinational that produces a range of discrete (versus integrated) semiconductor chips used in antilock brakes, cellular phones, televisions, and wireless computer networking. Discrete chips are somewhat more standardized than integrated chips, with competition based on low cost, high volume, customization, and speed to market.

The $110 million plant, opened in 1994, is located across nine hectares in a private EPZ in Laguna Province, approximately forty-five kilometers south of Manila. However, Discrete Manufacturing has been manufacturing chips and transistors in the Philippines since 1981, and the firm was constrained from creating an entirely "greenfield" plant in Laguna, primarily because it inherited much of the production and the entire unionized workforce from an older plant in Manila. And despite increased automation, most production is still relatively labor-intensive. The plant employs nearly thirty-five-hundred permanent workers, 85 percent of whom are factory workers. Workers earn an average of P11,622 or $305 per worker per month, which is 65–75 percent more than workers at the other two flexible production firms in this book. In 1999, the firm exported chips valued at more than US$500 million, making output per employee $8,211 or about the same as at Storage Ltd. but 65 percent less than at Integrated Production.

Flexibility, Quality, and Productivity Strategies

FLEXIBILITY Like other high-tech firms, Discrete Manufacturing has been introducing new, more programmable production machines for increased flexibility. But unlike Storage Ltd. and Integrated Production, Discrete Manufacturing cannot complement the new technology with the use of individualized, pay-for-performance schemes, forced overtime, or a twelve-hour workday since the unionized workforce has resisted these strategies through collective bargaining. Instead, the firm has begun experimenting with reorganizing production workers into teams. The ideas for the firm's reorganization come primarily from the Japanese lean production model, and all managers attend trainings on looking "beyond Taylorism" toward improving worker cooperation and problem-solving for continuous improvement. As the general manager wrote:

> The supplier which can consistently deliver the product or service faster, in any quantity the customer wants, anytime and anywhere in the world at the lowest

prices will win the market war. We saw this coming . . . We are gradually creating a nimble organization through self-directed teams and mini-companies. To do things faster, we need to make and implement decisions closer to the shopfloor. . . . Of course, doing things in a different way also requires different mind-set and skills. It requires different ways of allocating authorities and responsibilities. . . . We must throw away old concepts about power and authority and adopt new ways of working together.[14]

The General Manager's statement is instructive for a number of reasons. First, it acknowledges that technology alone cannot deliver flexibility and responsiveness. Second, it highlights the growing influence that customers have in defining how work is organized. Finally, it reveals that a central issue of reorganization is the battle over power, authority, and discipline on the shop floor. But as shown below, management's practices diverged significantly from its power-sharing rhetoric.

At the time of research, the company was introducing self-directed work teams (SDTs) in production, going much further in terms of shop floor reorganization than at either Storage Ltd. or Integrated Production. Officially, SDTs are part of the firm's high-performance strategy that *requires* worker participation in decision-making and flexible production. As the firm's promotional materials state, "Those close to the shop floor know the work intimately. They understand the problems and find solutions quicker. They can implement changes at a moment's notice. They can do away with bureaucracy." SDTs, which were still being piloted, were to be small—five to ten workers—and cross-functional, including supervisors, engineers, technicians, and production operators. The supervisor and engineer were to be "facilitators," while the team would elect an operator as team leader.

For example, one pilot SDT, in the molding section, consists of a supervisor, an engineer, and six production operators. While the supervisor plays a prominent role, the team leader is a team-elected operator, because, as one supervisor noted, "We are trying to promote operator-based management." The team had only been together three months, but had come up with production targets and behavioral goals and, according to the supervisor, had already achieved positive results, including higher productivity and higher quality.

However, while the company liberally espouses the rhetoric of worker empowerment, growing customer complaints and the need for more disciplined workers have been the real underlying drivers for work reorganization. Discrete Manufacturing maintains dedicated assembly lines for different customers, who come to the factory to audit "their" lines. After one such audit, a key customer complained to top management about "idle" operators and

14. Quoted from firm promotional materials.

a workforce that was "negligent and lacked discipline," which the customer felt was affecting product quality. The complaints about poor worker discipline have spurred the firm to improve operator performance, particularly as new automated machines were being introduced.

The firm's primary emphasis on motivation and increasing management control is evident in its selection of areas for reorganization. One of the first sections reorganized was a particular molding section, where chips are encapsulated in plastic. Unlike other sections in assembly, molding is considered "dirty" work, with fairly heavy and cumbersome machines that operate at high temperatures. For these reasons, operators and supervisors considered it a "male" area and all six operators were men (although the rest of the department is about 75 percent female). Ramon, the supervisor, considered the area problematic because of a high rate of absenteeism, and because some operators would not listen and had an "attitude problem." Workers in the molding section are also active union members and one is a shop steward.

Although the team did discuss production issues, the most evident changes were related to operator behavior. For example, one of the first issues the team took on was the high rate of absenteeism. Ramon remarked that previous negative individual sanctions had not improved the situation. But under the supervisor's guidance, the team set a collective goal and came up with a system of fines for fellow members. The shift from individual to team responsibility, and the added pressure of team members, seemed to have solved their absenteeism. Ramon also said that the "team orientation" of the group had improved cooperation. For example, when one operator had downtime now, rather than taking an "unofficial" break, he would often help another operator to meet the team's production quotas.

Workers' response to the SDTs were mixed. Jaime, a union shop steward and an SDT leader, commented: " It's good [*maganda*] because there's cooperation. We're able to know how to meet the demand. I give feedback as a [team] member and help brainstorm when there's troubleshooting. I like it better with SDTs than without." Similarly, Rene, also an active union member, said that with SDTs, "we've gotten a stronger voice to face and demand things from management." Yet others were less enthusiastic. One operator noted, "There's no real difference [between team and nonteam work]; with teams you just have to document things more. SDTs are really just one way of union busting."

Management's other key flexibility strategy is more in line with those of the other high-performance firms in this book: increasing automation. At about the same time they began introducing SDTs, management was also launching an ambitious new technical solution, the Manufacturing Innovation Line, or MI Line. The MI Line is quite similar to Integrated Production's automated A-Line. It is a fully automated continuous-flow assembly line, which takes whole silicon wafers through assembly to final testing and packaging on the same floor to minimize human handling. Because the lines are fully automated, operators on the line "mainly monitor the machines." But the new lines are also susceptible to frequent stoppage because when just one machine breaks down, the entire line grinds to a halt. Thus the new line requires swift action by trained operators or technicians, leading management to create a new position similar to the one at Integrated Production: the technical operator.

As at Integrated Production, the new position has the potential for adding a new rung on the promotional and skills ladder for operators. Yet Discrete Manufacturing has also deliberately chosen not to fill these new positions with existing line workers, who are overwhelmingly women, instead hiring all "fresh graduates" of either four-year engineering or two-year technical schools, all of whom happen to be men. When asked why all the technical operators were male, the production supervisors said, "We now have a hiring preference for males because of TPM (total productivity maintenance). The work is getting more technical and we have to compete globally." He hastened to add, however, that "women are not neglected. There is still a need for the expertise of women." When pressed what the nature of this expertise was, he listed dexterity, meticulousness, and patience.[15]

The new production process has not only resulted in a change of the gender of the operators, it has also radically reduced the number of workers needed. The MI Line requires only 4 operators a shift or a total of 16 MI Line operators. In contrast, the traditional production on an adjacent line that remains manual and semiautomated requires a total of 250 workers.

Union officers recognize that the technical operators are more skilled, have more say over where they work, and have less supervision. So rather than resist the introduction of the new positions, the union initially tried to get these positions filled by senior workers. However, management refuses to promote existing operators into these positions, arguing that current operators lack the technical education. The union has since become much more critical. They claim that the technical operators are overworked, forced to handle six machines at once, and only receive a small, performance-based bonus that can evaporate if there's a production breakdown. As one officer

15. It is interesting to note that at Discrete Manufacturing's more high-tech, integrated chips (IC) plant, the HR manager lamented the lack of female technicians and engineering graduates. He claimed that in advanced IC production, changes in technology and increased automation are increasing the demand for technical competencies and lowering the demand for manual dexterity. But he added, "There are still steps that require dexterity, which is no problem for females. . . . and patience is still needed because the parts they handle are small." The IC plant is currently working with several private training schools to train specifically female technicians.

put it, "It's just not worth it. There's more responsibility and much more work. The company's getting so much benefit [from multiskilling] but the worker gets only a tiny bit more."

But from the management's perspective, the MI Line has been quite successful and the staffing quite deliberate. The production supervisor stated that while they want to "promote operator-based management," they also want "to develop a new attitude with the new operators." The positions on the MI Line have much broader job descriptions and more responsibilities that are not covered in the union contract. The production supervisor continued, "There can be problems with members of the union, especially the old people. They can be very feisty. You can't force old timers to handle six machines. Most of them are complacent and you can't get them out of their comfort zone. . . . But with the 'fresh graduates' you can right away instill the right values. You can still intimidate them."

QUALITY   Quality is such a key concern that the leader of the Quality and Reliability program is a full manager, holding one of only nine top positions in the firm. Central to the quality program is ISO quality certification, which the firm, like other leading companies, views as absolutely essential to cultivating its customer base. But while workers have gained some formal skills through certification, the union also recognizes that the benefits of such a program are not evenly distributed. Lia, the union president commented,

> Things like ISO just put more pressure on workers. They need to document everything. Then the tracking records are used to monitor the workers and used for performance evaluation. . . . Productivity goes up but all the benefits for ISO go to the company. Workers only get "special awards" and token gifts, like free T-shirts or bags. . . . The company has been very successful in introducing ISO. But ISO *disorganizes* workers. It leads to individual competition between groups and between workers. It leads workers to think, "How can we beat the other groups?" But worker issues are never covered.

The firm makes extensive use of statistical process control (SPC) through the firm's shop floor control (SFC) system. Similar but slightly less automated than the manufacturing information systems at both Storage Ltd. and Integrated Production, the SFC combines automatically collected data from machines with inputs entered by operators. Operators must fill out "batch cards" that record the specific machines used, quantity produced, number of rejects, and yield. The batch cards themselves "travel" between workstation as a product flows through processing. Batch cards allow the firm to track the entire process to pinpoint the source of technical problems or bottlenecks. But as one supervisor noted, "The most important information on the batch card is the operator's name." Through SPC and the batch cards, the process engineers conduct daily and weekly checks of operator performance.

A final element of the quality program is the off-line quality teams. The five- to ten-member, cross-functional quality teams consist of operators, engineers, technicians, and supervisors. But nearly all the teams are led and organized by supervisors, engineers, or technicians. At the time of research, the firm had sixty-four active quality teams. The goals of the quality teams were focused on improving yield rates, reducing down time, reducing reject rates, and reducing material waste. Because engineers and technicians have the time and autonomy to troubleshoot, they are often the most active members and called on by supervisors to become team leaders. Thus while management often made inclusive statements such as "ops [operators] know better than us," the quality teams were clearly dominated by technical staff. The one exception during my research was a group of operators that had won the top award in a regional competition for "Best Productivity Jingle."

Indeed, many workers do not feel they participate as equal team members, are rarely listened to for ideas, and are instead used only to gather extra data, on top of their regular work and documentation demands. Thus while operators at Discrete Manufacturing have more opportunities to participate in off-line teams than workers at Storage Ltd. and Integrated Production, most still feel the programs are dominated by supervisors, technicians, and engineers, and participation remains low.

PRODUCTIVITY   Unlike most other firms, Discrete Manufacturing cannot simply ratchet up work hours or even use individualized, pay-for-performance schemes to boost productivity, since workers are protected by a collective bargaining agreement (CBA). There are three eight-hour shifts, and workers stated that there was no forced overtime or forced work on rest days. Instead, to boost productivity, the firm combines discipline, technical solutions, and teamwork.

One element that Discrete Manufacturing does share with Storage Ltd. and Integrated Production is its "visual management" system. A large, centrally located white board in each section displays the total output of each shift as well as the output of each machine and the output of each worker. Patricia, an operator of an automated wirebond machine stated,

> My target used to be thirty-two per hour, but now it's thirty-eight. If you're under target, they only ask you if there's a problem [you don't get a disciplinary action]. But . . . there is a big board where they can display that you are below target and you're even color-coded so that it's easy to see who hasn't hit the quota.

To combat productivity losses, the company has also begun a pilot total productivity maintenance (TPM) program. TPM is designed to include all levels of employees in the elimination of losses primarily through preventive maintenance. Workers who take part in cross-functional teams with engineers and technicians that meet both on the line and off-line. Workers are taught how to properly clean, lubricate, and inspect the machines and given diagnostic training to detect "abnormal conditions." For this, TPM workers were paid an additional twelve pesos per day. But like the SDTs, TPM also served management's labor control interest and supervisors targeted pilot machines and workers with discipline and performance problems. For example, the same "undisciplined" group in the molding section that was targeted for SDTs for its absenteeism, also had higher than average downtime and was thus targeted for TPM.

Negotiating Commitment: Human Resource Strategies and Labor Relations

Like the other high-tech firms in this book, Discrete Manufacturing tries hard to build worker commitment. But due to the countervailing power of the unionized workforce, the firm must offer mainly positive incentives and cooperative labor relations to boost workers' loyalty, effort, and attachment.

Management is keenly aware that many workers remain cautious, if not downright hostile, toward reorganized production. So to elicit worker loyalty, management "sells" its reorganization program as the embodiment of the firm's commitment to worker participation and autonomy. Yet their lofty claims collide with the firm's management-led TQM strategy, which seeks to limit, direct, and control work reorganization. In fact, the firm makes clear the bounded nature of the autonomy they offer: "Employees shall be empowered, encouraged, and supported *in their pursuit of customer delight*" (emphasis added). Studies show that providing real worker participation and autonomy is key to understanding workers' "enthusiastic compliance" (Hodson 1991). But worker participation in both the on- and off-line teams has been quite circumscribed, and once the projects are developed, workers are expected to follow strict "best practice" standards for the new processes, which actually diminish worker autonomy. Patricia, a diebond operator and an active union member, said:

> For me, team working is OK but we don't need SDTs. . . . because it only cuts down on their [management's] work. Instead of hiring new ones [supervisors], they make the operators do the job. Then, because it forces you to follow management's plan . . . it creates an illusion that workers are participating in management, but there's no increase in pay. They [management] have projects that you [workers] are forced to work on.

Another operator was similarly skeptical: "The way I look at it, only the bosses gain—the speed of the line is improved, the budget is shaved. But the gains go to the bosses, not the workers." Clearly, shop floor realities have not matched management rhetoric.

Still, workers do show interest in the reorganization, but for more instrumental reasons. First, they recognize that the creation of new positions (such as team leader) under SDTs opens up an additional promotional rung for older workers lacking the educational credentials to become technicians. For the union, the position represents the potential to usurp authority that once belonged only to the supervisor. In addition, while the union has previously resisted individualized pay-for-performance schemes, they have not resisted TPM because it provides new technical training, links these higher skills with higher pay, and allows senior operators to participate.

In terms of worker effort, the firm cannot rely simply on workers' voluntary efforts to boost productivity and is also deeply concerned with absenteeism. To improve job performance, Discrete Manufacturing relies primarily on "hard" TQM tools such as SPC, TPM, and traditional time-and-motion studies.

However, workers in general found the daily targets "fair," since they faced no forced overtime or disciplinary actions for not meeting targets. Workers in fact participate and take refuge in the time-and-motion studies as they set a clear standard for work and help guard against management's arbitrary ratcheting of production targets and quotas. While it is unusual that workers would accept the time-and-motion studies, the case of Discrete Manufacturing may be an instance of what Adler calls "democratic Taylorism" (Adler 1995). Under such a system, workers may accept strict rule and regulations if they have some say in how they are drafted and thus feel that they are "fair." This is also the basic underlying logic and reason for success behind the SDT in the molding section of setting their own absenteeism policy.

LABOR RELATIONS AND THE COLLECTIVE BARGAINING AGREEMENT (CBA)
Both the management and the union maintain that labor relations are cooperative and peaceful. The HR manager attributed the mutual trust and respect to management's efforts to maintain a nonadversarial HRM approach: "Union or no union, people have needs. If you can address those needs and both sides are transparent, then relations will be okay." He went even further, stating that managing a unionized workforce has many benefits: "We've had no problems with [the union]. . . . It's easier with a union, because we can talk to just one or two people, not two thousand. And a lot of people think it's more expensive with a union. But the wage bill here is twenty percent cheaper than at [three leading multinational electronics firms in the Philippines]. They have to buy people not to unionize."

While the union president agreed that labor relations are now generally cooperative, she attributed it not to management largess, but to an early show of union strength. She explained that many of the new programs (such as TQM) started in the early 1990s, when the current HR manager was hired. According to union president, the HR manager was brought in from the outside and had a reputation as a "union buster." "So we did a few mass actions to test him. In 1992, he took a hard line stance, but didn't bust the union. So he changed his tactics." The union has also kept up the pressure with job actions tied to collective bargaining negotiations. In 1998, in the run up to the negotiations, the union staged several actions, the largest of which was a boycott of the company's TQM day. This event, held on a Sunday, is largely a social function with free food and prizes. The union felt it was a ploy to "buy" worker consent to performance-based pay, which was for the first time to be negotiated in the CBA.

The management does try hard to maintain open lines of communication and has increasingly introduced strategic HRM practices similar to those used at Integrated Production. It surveys its employees annually, conducts an informal "coffee" with the general manager monthly for fifteen to twenty randomly selected employees, and twice a year department managers conduct a large open forum. While operators appreciated the opportunity for communication, they were also wary. Shala, an operator, stated that although the management-sponsored forums were "an effective venue," she also said that they are "one way for the boss to get to know each and every thing that happens." Another noted that operators "are consulted for formality's sake but management only pushes whatever plan they have."

The main arena for negotiation is not the management-sponsored communication forums, but the collective bargaining sessions. The CBA is the centerpiece of union strategy to provide for its members, while for management, its their opportunity to win worker cooperation and to define and formalize "management prerogative." Explicit in the CBA is a long list of areas that management deems outside arbitration with the union, including work assignment, performance standards, and rules on hiring, hours of work, and safety. Management has also included in the latest contract an agreement for the union to "support the management initiated Performance Incentive Program." According to the mangers and supervisors, the union has not formally demanded more say in production matters or more decision-making power. For example, the TQM manager noted, "The CBA is purely benefits focused. They [the union] have negotiated all kinds of leaves and social support." A supervisor and former union member also commented that "the union doesn't have input in reorganizing work. That's management prerogative. [The union] concentrate[s] on wages."

Union representatives for the most part confirmed management views of union interest. While the union did want to influence the new work schemes informally, such as taking leadership roles, they did not want to get too involved in the "business of management." Rather, the union has focused on boosting base pay and benefits, particularly ones that supported employees' nonwork lives. Many of these, which broadened the workers' safety net, are not found in other, nonunionized firms.

The core of this safety net is seniority-based pay. While starting pay is P212 per day, or just 6 percent above the minimum wage, the salary range is quite high. A union member with the most seniority might earn up to P19,500 per month, or slightly less than four times the starting rate. The average monthly wage for all workers in the bargaining unit, which includes technicians and QA inspectors, is P10,983 per month or about P477 per day. This is above what labor unions and advocacy groups considered a "living wage" for a family of six. Discrete Manufacturing workers also work only twenty-three days a month, versus twenty-six days a month or more at many other firms like Storage Ltd. and receive non-performance-based bonuses for years of service and holidays.

The union has also been able to negotiate an impressive benefits package. Union members enjoy seniority-based protection against layoffs and subcontracting; full family medical and dental insurance; and a diverse set of leaves, including eighteen days of vacation leave and nineteen days of sick leave a year, six weeks paid maternity leave, seven days paternity leave, three days of emergency leave, three days bereavement leave, and a day off on their birthday. Similarly, they have negotiated an array of loan and subsidy programs—such as emergency, education, and housing loans, and subsidies for education, medical costs, childbirth, transport, bereavement, and food expenses.

It is important to note that these benefits—which exceeded those at any of the other twenty firms visited during field research—were *negotiated* and reflect workers', and particularly women workers' needs and priorities. While Integrated Production did provide some generous items such as a free computer, such benefits were entirely decided on by management. And unlike at other firms where bonuses were based on individual-, team-, and firm performance, Discrete Manufacturing workers have negotiated seniority-based pay and job security, giving them more protection from market and performance fluctuations.

Finally, workers also enjoy seniority-based promotions. The company offers a number of internal career ladders, both technical and nontechnical, and provides training and/or funding for training to help workers gain the skills for higher positions. Unique among the firms in this book, a job ladder has been constructed from operator to technician (three grades) to engineer (three grades). For example, through the firm's college scholarship program, Anna, a production operator, was able to return to school, earn a BS degree in computer science, then get promoted to a technical position as a junior production specialist. Operators can also climb a nontechnical ladder and, indeed, most of the supervisors interviewed were former operators. Others, such as Victoria, have gone even further. Victoria began as an operator in 1989, became a line leader, then supervisor, and is currently in charge of the TQM program for an entire production department. The HR manager confirmed that at least 60 percent of supervisors and managers were recruited from the internal job market.

In sum, although neither labor nor management really believes the "win-win," "empowerment" rhetoric of flexible production advocates, both sides recognize the stakes and potential advantages and each has contributed to a broadly based and relatively stable form of "bargaining commitment." In particular, the hard-won package of decent pay, good benefits, and security of tenure has generated a quite high level of worker attachment to their jobs and overall commitment. Still, workers' loyalties remain split between the firm that makes it all possible and the union that has negotiated the spoils. This issue of split loyalties and workers' own attitudes and responses will be taken up in chapter 5.

Variation Explained

The firms presented here show clearly that although flexibility is necessary for advanced manufacturing, it is not all of a piece. Allied Power, for example, practices only a kind of defensive numerical flexibility, extending work hours and sweating labor when demand is high, and simply cutting back on hours and workers when orders fall. The remaining three firms all pursue more dynamic internal flexibilities, yet their production and labor control strategies diverge significantly.

The patterns vary for a number of important reasons. First, at a macro level, the Philippines experienced a debilitating debt crisis in the 1980s followed by structural adjustment packages attached to International Monetary Fund loans that molded available state strategies to promote economic development. These changes were outlined in chapter 2. By the mid 1990s, the Philippine government had embraced neoliberal economic prescriptions for growth, substantially deregulating foreign investment and employment, particularly in the new, privatized export processing zones (World Bank 1997). This produced an environment that allows for and tolerates many different employment strategies at the firm level, despite relatively strict national labor laws on the books. In addition, as argued in the previous chapter, there have been important changes in the nature of global competition itself (Borrus et al. 2000). The multiple and contradictory demands of competitive performance mean there remains much confusion and experimentation over just what kind of workplace changes are actually necessary to lower costs, raise productivity, increase quality, and cut ramp-up speed to mass customization.

Focusing more closely on the firms themselves, three other sources of variation become clearer, with the first two being the most crucial. First and foremost, all four firms' major products are in different subsectors within a quite diverse industry. Thus differences in the competitive character of their respective markets, the technological intensity of their products, and the labor and technical requirements of their production processes all play key roles in how work is broadly organized. For example, both Power Tech and Allied Ltd. produce low-technology standardized consumer products using mainly older machinery. They thus compete almost entirely on price. As such, the firms, particularly Allied, keep wages low to shave costs and invest little in improving product quality. Both firms rely instead on numerical flexibility—adjusting the *amount* of labor applied in production—without significantly changing the actual production process. With low wages and an uncomplicated labor process, management focuses on worker effort using simple direct and coercive control over workers, as there is little need to upgrade skills or promote deeper loyalty. For Power Tech, it was only after the firm began moving higher up the value chain and purchasing more automated, surface-mount technology that management began paying attention to international quality certifications and shifting away from an entirely despotic approach to labor control toward the use of more positive incentives.

The three remaining firms produce technologically more sophisticated electronics requiring more stability and quality in production, and thus cannot depend entirely on such coercive strategies. However, their products and production processes are still significantly different, leading to divergent work regimes. Storage Ltd.'s hard disk drives, for example, are more standardized products than either discrete or integrated semiconductor chips. As such, high-volume, low-cost production is a competitive necessity for HDDs. However, HDDs also have far more moving parts than semiconductors and much of the production process has not been automated. With pressures to cut cost, yet with automation less of an option, Storage Ltd., as other HDD firms, must minimize labor costs while maximizing productivity (McKendrick et al. 2000). Storage Ltd.'s labor force of more than nine thousand workers is two and half times that of Discrete Manufacturing, and nearly seven times that of Integrated Production. The acute technical and complex market demands of HDD production, coupled with its labor-intensity at the assembly and testing stage, have led Storage Ltd. toward a "disciplinary management" strategy that focuses on maximizing productivity and securing quality though surveillance-led control, while still keeping wages low. As we will see in the next chapter, securing worker commitment and keeping turnover low under such extreme conditions requires an equally intrusive localization strategy into workers' communities and lives outside the factory.

For Integrated Production, the more capital-intensive production of high-quality, customized chips has led the firm toward more automation and thus a reduced emphasis on labor costs. Rather, with expensive machinery, high-cost products, and a much smaller labor force, the firm has chosen to avoid the potential risks of a coercive strategy and instead can afford to focus on an internal strategic HR approach that stresses more positive incentives. To a lesser extent, these technological pressures are also at work at Discrete Manufacturing, although management there has been more constrained in their strategic choices.

The clear differences between the work regime at Discrete Manufacturing and the other firms point to a second important source of variation: the level of workers' *collective* negotiating power to resist and contain management's flexibility strategies. At Discrete Manufacturing, the unionized workforce has successfully limited management's ability to entirely dictate how work is organized and how worker commitment should be secured. Discrete Manufacturing does share some key characteristics with Integrated Production: both produce customized semiconductors, are introducing similar automation strategies, and are relatively less labor intensive than Storage Ltd. Both have also used more positive incentives to motivate workers. However, at Discrete Manufacturing, the package of benefits has not been the product of company largess but has been collectively negotiated by workers. Recall that the HR manager, while seemingly tolerant and accepting of collective worker representation, was initially identified as a "union buster." It was only when that strategy failed that the firm shifted to a more cooperative stance. This preference for a nonunion environment is also clearly demonstrated at Discrete Manufacturing's entirely new plant—discussed in the next chapter—which produces integrated chips just a few miles away. While management has long trumpeted its good working relations with the union, it has taken many steps to ensure that the new plant remains union-free.

Third, the firms are all subsidiaries of multinational firms headquartered in different countries. It is not possible to generalize about national differences regarding work organization and localization strategies from single case studies. But the four firms do exhibit divergent or "embedded" organizational tendencies not entirely explained by technological differences (Dicken 2003; Whitley 1999; Dunning 1993). For example, Discrete Manufacturing's union acceptance may reflect its European parent corporation's greater openness toward and experience with a unionized electronics workforce. When asked about their distinctiveness as one of the only unionized high-tech firms in the Philippines, the European general manager responded, "Look, in my country, if you have five workers together, you already have a union. We don't necessarily like it, but we can deal with it." This attitude is a far cry from statements made at either Storage Ltd. or Integrated Production, both of which were vehemently antiunion. Similarly, Integrated Production has patterned its human resource and labor relations policies after its parent company in California, where the electronics industry has used an HRM strategy and closely follows what Milkman (1991) has called the American nonunion model (see also Pellow and Park 2002; Chun 2001; Hossfeld 1995). Finally, Storage Ltd., while clearly driven primarily by technological factors, nevertheless also displays many of the characteristics of Japanese-style lean production and intense control found in other Japanese electronics firms, not only in the Philippines but also in Europe and along the U.S.–Mexican border (Delbridge 1998; Kenney et al. 1998; Chow and Hsung 2002).

Flexible Accumulation and the Expansion of Management Prerogative

Despite the variation in work regimes, firms practicing high-tech flexibility do exhibit a number of commonalities in how they organize production. First, in a curious way, workers at the least flexible and lowest-tech firm, Allied Ltd., have been able to carve out a small space of autonomy as long as they meet their quotas. Unlike at Storage Ltd., the inattention to quality at Allied gives workers a small respite from management control. As one operator commented, "You just do your work. As long as you finish the work, they leave you alone." But as the high-performance literature stresses, and the three higher-tech firms demonstrate, coercion and numerical flexibility is insufficient for advanced competitiveness, and the shift to teamwork can improve autonomy and discretion over one's work (Appelbaum et al. 2000; Kenney and Florida 1993). However, contrary to the literature, flexibility and quality in the firms studied have generally been achieved by workers other than production operators. In the assembly and testing segments of electronics manufacturing, engineers and technicians dominate total productivity maintenance (TPM) teams, quality teams, and even self-directed work teams, and are able to exercise some autonomy and control over their work. One junior quality engineer explained,

> I have a lot of control. I can come up with my own improvement proposals. We discuss with the managers and if there is no contention, we carry it out with the techs [technicians] and ops [operators] . . . I never had a proposal rejected. And it's not just one aspect of production. We can have input on the entire process. . . . In morning meetings, he [supervisor] tells us what the directors want, then we [engineers] meet later by ourselves to discuss improvement. In a month, I'd submit about three proposals.

She also noted, "For ops, they don't have any autonomy, we do everything for them. We design their jobs and their daily routines. And we have high quotas, say 100,000 HDDs. We [engineers] just monitor the process and inform other departments. But it's the ops that have to produce the 100,000." When asked why the engineers bothered to give proposals for improving production—in essence their discretionary effort—the engineer replied, "My supervisor had high regard for me, and that's what mattered. Respect. The most important thing was my boss listened to me." She also cited good relations between the Japanese and Filipinos, high overtime pay, good benefits, and the prospect of being sent to Japan for further technical training.

Thus rather than production worker input, the more complex technologies seem to require participation from those with formal engineering and technical education to troubleshoot and cope with the increasingly technical nature of production.

Second, there seems to be clear evidence that complex market pressures have intensified control over production, leading to *greater* standardization and quantification (through statistical process control) of production line work. Storage Ltd., for example, has minimized its use of external flexibility such as layoffs and the use of contract labor. Yet many of the internal strategies, such as forced overtime, focus on work intensification rather than actually reorganizing production or changing authority relations. For shop floor operators, rather than empowerment and an expansion of worker discretion, the introduction of new technologies has primarily meant subjugation to higher levels of technical and bureaucratic control (Tsutsui 1998). This finding is similar to conclusions drawn by Jane Collins (2003) in her study of subcontracted apparel production in Mexico and by Steven Vallas (2003) in his research on technological change and team work in modernizing pulp and paper mills (see also Vallas and Beck 1996). As in these studies, the introduction of higher technologies, such as automated manufacturing information systems, has reduced, rather than expanded, the roles played by production operators. Line automation and computerization have also reduced the need for some operator skills, like soldering, and *narrowed* previous participatory roles, like manually collecting statistical process control data.

But if competitive performance systems do not require greater discretion from operators, they do require more lengthy and formal trainings, due to demands of ISO 9002 certifications and the needs for a stable, reliable workforce that will not jeopardize or interrupt production. Thus, as Delbridge (1998) found in his study of a Japanese consumer electronics firm in southern England, the emphasis on marginalizing uncertainty in production also means an increase or consolidation of management prerogative over the workplace and a more disciplined workforce. But discipline can be achieved in a variety of ways. Power Tech disciplined its labor by simply downsizing a large portion of the workforce, busting the fledgling union, and bringing the remaining workers "back to harness." Storage Ltd. also chose a "disciplinary management" path in which control over workers remains quite intense and workers are subject to heavy surveillance in all areas of the plant. Even at Integrated Production, which chose to manage mainly by incentive, workers were subject to intense scrutiny on the job and were not given the opportunity to participate substantively in decision-making in production. Finally, even in the relatively anomalous case of unionized Discrete Manufacturing, management consistently pursued a strategy to preserve and expand its prerogative, first through collective bargaining to gain the voluntary cooperation of workers, and also through new forms of work organization.

Thus a clear trend across all four case studies has been the expansion of management prerogative, though manifest in different forms in different companies, as competition has become more intense and assembly and testing manufacturing more integral to the parent firms' accumulation strategies.

New Technologies and Gender Stratification

The three flexible high-tech firms also share another similar effect of technological change: persistent gender stratification. Storage Ltd., with the most labor-intensive labor process of the three firms, is also the most gender stratified. Because it recruits women operators almost exclusively, yet creates a promotional ceiling for production workers, the firm maintains and reproduces a rigid gender segmentation and hierarchy within the plant. Gender stratification is also clearly evident in the other two higher-tech firms. There, although technological upgrading has the *potential* to improve conditions and opportunities for women operators, in reality it seems that automation is likely only to deepen inequalities. The emergence of the technical operator position at both Discrete Manufacturing and Integrated Production would seem to provide an opportunity to add a crucial link across job ladders. Other studies have also pointed out the potential advantages that technological upgrading might have, particularly for women production workers (Chow and Hsung 2002). But in large part because the firms choose not to promote current (women) operators into these positions and instead fill them with fresh graduates of technical schools (men), the potential for mobility remains unrealized. Even at Discrete Manufacturing, where the union attempted to win these positions for current operators, "management prerogative" in matters of plant restructuring and job assignment thwarted union efforts. Hiring only male technical operators represents a lost opportunity to promote women's occupational mobility and demystify gendered stereotypes regarding women's relation to high technology.

This ongoing gender segmentation amidst industrial upgrading has important implications, particularly for wider occupational stratification, women's manufacturing employment, and the gender wage gap. As Mehra and Gammage (1999) note, for many developing countries, in the early phases of industrialization, gender inequality in labor market participation initially declines. This is also true for the Philippines, which has experienced a "feminization" of manufacturing since the 1970s because leading employing sectors such as electronics and garments have so far preferred women. The greater employment of women in manufacturing has helped close the wage gap between men and women and reduced women's unemployment below the level of men's unemployment.[16]

However, the trend from a number of other developing countries, such as Hong Kong and Singapore, which have already experienced the kind of industrial restructuring that the Philippines electronics industry is only now undergoing, is not encouraging. In these countries, technological upgrading has led to increased sectoral and occupational segmentation by sex (Joekes and Weston 1994). Consequently, as sectoral hierarchies and skill differentials became increasingly pronounced, the wage gap has again widened as women's wages fall farther behind men's (Mehra and Gammage 1999).

Not only is there a widening divide between men and women, but an actual drop in the demand of women workers as technological upgrading spreads. In South Korea, for example, as a wide variety of industries became increasingly capital-intensive and the total number of production workers fell, women workers were shed faster than men (Mehra and Gammage 1999). Similar trends have been noted in Malaysia (Ng and Mohamad 1997) and Taiwan (Berlik 2000).

Clearly, this "defeminization" of the workforce as production jobs become fewer and increasingly technical is evident at both of Discrete Manufacturing's plants and at Integrated Production. Integrated Production, with the most automated plants in this book, is also the most capital-intensive, has the highest percentage of technical workers, and overall the most balanced gender ratio. Indeed, the HR manager was right when he noted that at Integrated Production, "guys were beginning to gain ground." But as their workforce was becoming increasingly dominated by men, they sought to "masculinize" production, emphasizing the problems with women workers (who get "sick" or pregnant) and reworking the gendered stereotypes of the "new" assembly work, which now required more "stamina" and "strength": in other words, work more appropriate for men. At both plants, automation on the most advanced lines were threatening to remove all human handling of individual chips and therefore likely to lead to downsizing of women production workers. If the trend continues, it may signal a "masculinization" of advanced electronics manufacturing and suggest that women's employment gains in the industry may be short-lived.

In many ways, the manipulation and gendering of the new technical operator position exemplifies another facet of what Elson (1999) calls the "gen-

---

16. It should also be noted that the reduction in the wage gap between women and men may also be due to the erosion of men's position and a deterioration of some men's work because of (for example, in construction and transportation) overall restructuring and the shift of the economy toward services (Armstrong 1996).

der paradox of restructuring." Women's individual bargaining power may be increased by such technical training and the opportunity for formal sector employment, yet their extremely low bargaining position ensures that they remain more and more at the mercy of management, market forces, and technological change.

4 *Strategic Localization and Manufacturing Commitment*

NANCY, a production worker at the Cavite Business Park, described her days at an electronics factory: "A friend told me there was an opening . . . , so I lied about my age . . . and they accepted me, by the grace of God. . . . But they make us work for twelve hours without sitting down and they don't consider how tired we get. Sometimes we ask them if we can rest, and they say, 'Well, you have to be pregnant for you to sit down and rest.' All the higher ups, from technician on up, they're sitting. They are all sitting down, they are all supervisors, they are all men. If they catch you sitting down, they always ask, 'Are you pregnant? Before you can sit there, you have to get yourself pregnant first.' It's horrible. You get so humiliated. . . .

"A lot of us collapse on the job because we're so tired. We used to have three shifts: six to two [o'clock], two to ten, and ten to six. Then they compressed it to six to six [for both shifts]. We are so tired. A lot of people have aching legs. If something is dropped [on the floor] we have to let them know before we can pick it up. If there is a supervisor around, we have to ask them to pick it up, then we can throw it in the red box. If no one is watching, we pick it up; we just pick it up to relieve the pain in our legs. And we pretend to pick something up just to relieve our legs. Many don't eat during break time so they can sleep. Most often, in one day, three people collapse, sometimes two people or just one. But yes, everyday. It's really hard, it's terrible. We only get thirty minutes break time [for lunch] now. Fifteen minutes [morning break], thirty minutes [lunch break], then fifteen minutes [afternoon break]."

Interviewer: "So, do you want to keep working here?"

"Well, yes it's okay."

Interviewer: "Okay? really?"

"Well, my father lost his job and if I wasn't married and didn't have a job, we'd die of hunger. If I hadn't married, I'd still be going to school. That is, if my father still had a job. But things were so hard. I work now just so we can eat and drink. Right after my father lost his job, sometimes we only ate twice a day. So I came here, to help my father. . . .

"It's hard right now, but I'm also a regular [permanent employee]. It's so hard to find another job and other jobs are only contractual, maybe I should just remain here. Besides, I'm too old now to get another job like this [in electronics.]"

Interviewer: "How old are you?"

"Twenty-one."

Nancy's painful yet common predicament illustrates a key element that many workplace studies often neglect: that consent—why Nancy would consider such a punishing job "okay"—hinges critically on the availability of a worker's options and (in)ability to walk away (Burawoy 1979; Hodson 1997). To show just how important local conditions are for flexible accumulation, this chapter ventures beyond the high-tech factory, examining how firms tap into the broader foundations of consent and reach deep into worker communities and local labor markets to help manufacture stability and worker commitment. Specifically, these processes, which I call strategic localization, involve three primary tacks: selective and gendered recruiting, suppressing or preempting union organizing, and conspiring with state officials to regulate labor and manage the export processing zones (EPZs). Detailing how the four high-tech firms strategically localize their production in line with their respective work regimes helps better explain the varied types of worker commitment the firms generate: low commitment at Allied-Power, alienative or "coerced" commitment at Storage Ltd., "purchased" commitment at Integrated Production, and "bargained" commitment at Discrete Manufacturing.

Strategic localization, however, involves more than just firms and their workers. As discussed in chapter 2, the Philippine state has played an important and increasingly nuanced role in luring advanced electronics manufacturing into its expanding EPZ program. In addition to the burgeoning of zones from four to sixty-five in just eight years, there have also been important changes in the character of state intervention. Both nationally and locally, the Philippine regulatory state has rolled out a new set of neoliberal institutions to provide the "proper" conditions for flexible accumulation (Peck and Tickell 2002). These changes include the selective reregulation of local labor markets, new labor control tactics, and the reorganization of the zones themselves.

The developments in the Philippine EPZ program are chronicled below, beginning with the transitions from the first public zone in Bataan, to the modified public zone where the Allied-Power group is located, to the privatized zones where the three higher-tech firms in the book are found. The changes in zone organization are integral to the wider changes in the political apparatus of production and work restructuring. Thus the evolving zones are presented along with the case studies of the four firms and their extra-factory strategies, demonstrating the breadth of local labor control regimes and how local conditions have become only more important.

Recipe for Conflict: The Bataan Export Processing Zone

Portraits of "peripheral" manufacturing, particularly in EPZs, have often focused squarely on cheap labor, shop floor coercion, and direct state repression. The EPZs themselves are typically depicted as little more than detached sweatshop enclaves for mobile manufacturers with few local linkages (Elson and Pearson 1981; Fernandez-Kelly 1983; Amirahmadi and Wu 1995; Klein 2000). This is also how Philippine EPZs have been presented since the opening of the first zone in 1972 (Ken Ohara 1977; Aldana 1989; Perez 1998). But this static characterization masks important changes, both in the development of the zone program and in the nature of production within the zones.

The Philippine government began its strategy of export promotion through EPZs in 1969, passing legislation (Republic Act No. 5490) creating the Foreign Trade Zone Authority. By 1970, work began at the site of the first EPZ in Mariveles, an isolated fishing village at the southern tip of the Bataan Peninsula, at the strategic mouth of Manila Bay. But it was not until 1972, when President Ferdinand Marcos declared martial law, abolished Congress, and made nontraditional exports the engine of his growth and industrialization strategy, that sufficient foreign funds were borrowed to complete the construction of the Bataan EPZ (BEPZ) (Diokno 1989). Within two months of declaring martial law, President Marcos ordered Presidential Decree No. 66, which created the Export Processing Zone Authority (EPZA). The decree gave the new national agency the authority to evict the residents of Mariveles living on the zone site and broad policing powers within the zone, superceding the regulatory laws of local or provincial governments. The zone, which occupies 345 hectares, was designed to host eighty firms and projected to employ some sixty thousand workers at full capacity. Like all supplies and equipment, workers too had to be either shipped or trucked in. Most workers were young, single women coming from nearby provinces such as Pampanga and Tarlac. Critically, the government chose to house workers in large, concentrated dormitories adjacent to the zone, in part because the town's isolation provided no existing infrastructure for so many people. The zone and housing scheme thus isolated workers and made them dependent on government-subsidized housing, while also making worker surveillance and control easier and more centralized.

By 1976, twenty-nine firms had set up production, mostly sewing garments and assembling electronics. By 1980, the zone employed 28,260 workers, more than 80 percent of whom were women. These women workers were young (97 percent below age thirty), single (81 percent), and relatively well educated (more than half with high school certificates and 30 percent having gone through college) (National Commission on the Role of Filipino Women 1985).

But working conditions in the zone were poor, particularly because many firms used casual labor and paid less than the minimum wage (Scipes 1996). And beginning in late 1980, in the face of slowing world demand, many of the firms experienced production slowdowns, shutdowns, and layoffs. From November 1980 to September 1981, one-fifth of the entire zone workforce was either laid off or on forced vacation leave and in 1983, more than two thousand workers lost their jobs entirely (Diokno 1989). The economic slowdowns and worsening conditions in the zone coincided with a rise in labor militancy in the late 1970s. The rise in militant unionism grew in response to heavy labor repression under martial law and as part of the growing movement against the Marcos regime (Scipes 1996). The new labor upsurge culminated in the founding of the openly militant Kilusang Mayo Uno (KMU) labor alliance on May 1, 1980 (West 1997). KMU member unions and other more moderate labor groups had already begun targeting BEPZ for organizing, and after the partial lifting of marital law in 1981, organizing in the zone exploded (Scipes 1996).

Despite heavy repression by the state and zone authorities, the isolation and heavy concentration of workers in common housing blocks proved a tinderbox for union organizing (Vasquez 1987). Following a string of individual strikes in the zone that were met with police violence and the arrest of union leaders, the KMU organized a zonewide sympathy strike in June 1982 involving thirteen thousand workers.[1] The strike, which was the first general strike in the Philippines following the lifting of martial law, shut down the entire zone for four days. The strike was also noteworthy for being the first zonewide general strike in an EPZ anywhere in the world (Dejillas 1994; Scipes 1996).

Although the Marcos regime retaliated by imposing a strike ban in the semiconductor industry, the first BEPZ general strike led to the founding of the Alyansa ng Manggagawa sa Bataan—Bataan Labor Alliance (AMBABALA), and in 1983 a wave of some thirty-one strikes in twenty-two firms. Organizing in the zone continued and at its height in 1987, 89 percent of the 18,224 workers in the zone were organized into 41 unions (Anonuevo 1994; Perez 1998). The high level of organization is remarkable, considering that overall unionization rates in the Philippines were only 20 percent and no

---

1. Scipes (1996) claims twenty-six thousand workers, or the entire zone workforce, participated in the strike, since the entire zone was shut down.

unions existed at the other two public zones in operation at that time, in Baguio City and Mactan (Aganon et al. 1997).

The hard lessons learned at the Bataan zone were not lost on the central government or the central zone authority. In 1981, before the BEPZ strike wave exploded, the government had planned to set up twelve additional public zones. But in part because of the strike wave at BEPZ, considered disastrous by the government, the zone authority decided to push ahead with only one other zone: the Cavite Export Processing Zone (CEPZ), later renamed the Cavite Economic Zone (CEZ).

"A Cheapskate's Paradise": Allied-Power and the Cavite Economic Zone

Pawning ATM cards at usurious "5–6" interest rates in order to survive [in the "5–6" interest rate scheme, for every five pesos borrowed, six pesos must be paid back, a 182 percent yearly interest rate]. Delayed and below-minimum wages. A perennial diet of instant noodles. Sleeping in hot, crowded quarters. Forced overtime work or "OTTY" (overtime, thank you). "Finish contract" every five months. Perpetual contractuals. Union-busting, summary dismissals. Runaway shops. No union, no strike policy.

These conditions rule the daily lives of some 48,000 workers at the Cavite Export Processing Zone. . . .

If the CEPZ is not a paradise for workers, it is certainly one for the foreign and domestic capitalists who manufacture their goods here. In fact, it is a cheapskate's paradise. Filipino workers slave away for so little money, and brown sweaty bodies are offered cheap on the altar of the global market.

Ma. Ceres P. Doyo, *Philippine Daily Inquirer*, May 1, 2000

What used to be a sleepy coastal fishing town is now a bustling strip of unchecked urbanization along the narrow highway. It's shift change and traffic is at a standstill. Young women, still in their uniforms, pour out of the zone and into the streets. They're threading their way back to their makeshift boarding houses, which are tacked onto the older homes around the zone. A near solid wall of food stalls, sari-sari stores, motorized tricycle stands, shops, and a few Korean and Japanese restaurants line the road. I'm eating lunch directly across from the main zone gates but the newish restaurant I'm in is nearly empty. They serve the usual fare: fried rice with sausages, shrimp paste rice, tilapia, etc. I had veggies and a fish over rice. It cost P60 ($1.50). I asked the owner why it was so empty and who eats here. She told me it's mainly "foreigners" from the zone or "walk-ins" from the local community. Workers in the zone, she said, can't afford fish.

Field notes, 30 April 1999, Rosario

The Provincial Labor Market and Allied-Power's Gendered Recruiting Strategies

As demonstrated in the previous chapter, low-tech electronics firms such as Allied-Power, like their counterparts in Bataan, continue to pay low wages and rely on direct shop floor control. But to help curb labor militancy, Allied-Power has also buttressed their shop floor control by expanding their labor control strategies to other domains outside the factory. Before tracing Allied-Power's external strategies, I briefly sketch a picture of the Cavite labor market to show the social context in which production and recruitment takes place.

Cavite, the province just south of the National Capital Region of Metro Manila, has a population of 1.6 million and a rapid yearly population growth rate of 7 percent, due mainly to high in-migration. Eighty percent of the population lives in urban areas, where, interestingly, women outnumber men. The sex ratio is most extreme in Rosario, where the CEZ is located: for every hundred women there are only ninety-five men, reflecting the high level of female in-migration. It is the opposite case in rural areas, where men now outnumber women. The overall unemployment rate in Cavite is 10.2 percent. But this relatively low rate masks the high level of underemployment, which stands at 18.5 percent. A majority of those in the formal sector, 55 percent, are employed primarily in the low-paying service sector, while industry employs 31 percent. Employment in agriculture has dropped to 13 percent in this once, agriculture-based province. Overall, despite the mushrooming of industrial estates, more than 25 percent of all households fell below the official poverty line (Province of Cavite 1999 and NSO data).

The general labor market conditions ensure a steady stream of young applicants into CEZ and, as Peck argues: "One reason that secondary work exists is the *prior* existence of a group of workers who can be exploited in this way" (Peck 1996:69, emphasis in the original). Indeed, both Power Tech and Allied Ltd.—with their low-tech production and cost-minimizing strategies—strategically recruit workers who would be willing to accept below-average wages and poor working conditions. Reflecting and (re)producing a pattern that has been amply documented in export zones around the world, firms like Power Tech and Allied have sought a young, single, inexperienced, and female workforce (Fernandez-Kelly 1983; Ong 1987; Chant and McIlwaine 1995; Freeman 2000; Erdralin 2001; Wright 2001; Chow and Lyter 2002; Salzinger 2003; Mills 2003). HR managers, zone administrators, and the head of the Industrial Relations Division (IRD) who helps screen zone workers have all drawn on the gendered stereotypes typical in EPZs worldwide to justify their preference for such a workforce. For example, one zone administrator commented that electronics firms have come to the Philippines because Filipinos are more dexterous and agile and that firms prefer young women workers because women have more patience: "Women can still do it after several hours while men get restless." Similarly, the IRD chief claimed that firms sought young workers because they had better eyes and dexterity and recruited women because they were more "industrious" and "patient" then men.

While Power Tech and Allied differ in their recruiting and screening processes, both were successful in building young, female workforces, due in large part to the high levels of general un- and underemployment in the local labor market. As a sub-subcontractor offering only low pay and unstable production, Allied's strategy has been to hire very young women, some even below the legal minimum age for work. Like many firms in the CEZ, Allied does not have to formally advertise when it has openings, it simply informs its current workers and puts up a sign outside the factory. The zone administration's Industrial Relations Division also maintains a file of thousands of prospective workers who have come to the zone in search of work. One Allied worker noted, "I just walked and walked and I saw the sign. I saw people all lined up, so I got in line as well."

Angel, who started in 1992 and came originally from the province of Leyte, describes how she ended up at Allied:

> After finishing high school, I went to Manila to look for work. But in Manila, you need to be eighteen. I had been waiting from a call from Uniden, in Manila, but they couldn't hire me because I wasn't eighteen yet. Then a friend from church said Allied was hiring and that they need young ones and I was only seventeen. There were about one hundred of us. But only twenty-three of us were hired.

When Allied first started production, there were very few requirements and the screening was almost arbitrary. Angel continued her story about the hiring process:

> We had no exams. We were personally approached by a Korean, Mr. Park, who interviewed me. He asked me for my bio-data, then asked me to stand. Then he asked me to turn around. Then asked me to count from one to fifty like this [counting on her fingers]. It should be to the end. He's watching your hands, how fast you are. It seems if your hands are fast, you will work faster too.
> *Interviewer: why did he ask you to turn around?*
> I don't know, I think it's your personality. They say Koreans are like that, they look at your personality. I guess it's part of your personality.

Later, the firm instituted a more systematic screening process. One worker, hired in 1995, explains:

There were a lot of exams, and each for only fifteen minutes. In fifteen minutes, there are more than a hundred items. Algebra, arithmetic, abstract problem-solving, and essay. . . . We were one hundred and twenty-four, only sixty-four passed. Afterwards, we had a technical interview. They tell you all the company policies and then ask if you agree and if you will be willing to do overtime or work night shift. The shifts were eight to eight, twelve hours. You need to answer that you're willing to do it and when they ask if there is anything else you need to answer "nothing."

The workforce at Allied closely reflects management's ideal hiring criteria and demonstrates the strict vertical and horizontal gendered division of labor: all five of the foreign managers are male, while all local office support staff are female, as are 80 percent of the production workers. For operators, nearly all are only high school graduates.

At Power Tech, the recruiting and current worker profiles are similar to that of Allied. Power Tech seeks to recruit production operators who are female, eighteen to twenty-two years of age, single, and high school graduates. But the HR manager added that they did make exceptions to the high school graduation requirement and that there were some hires who had lied. Meanwhile, when hiring professional or technical employees, the firm sought males over twenty years of age with at least two to three years of vocational education and/or some engineering classes. The HR manager said that they fill most positions through "free sources," such as employee referrals, word of mouth, and postings in front of the factory. Occasionally, the firm also uses the local government-run Public Employment Service Office (PESO) to screen local applicants for jobs. Power Tech's workforce also reflects their hiring preferences. Eighty-eight percent of their production operators are female, 80 percent are high school graduates, and 16 percent have some college education.

In sum, both firms relied heavily on (and helped reproduce) gendered stereotypes to justify their hiring preferences and, because of slack local labor markets, were easily able to build almost entirely "feminized" workforces. Whether or not the management actually believed their own reasons for hiring young single women, they nonetheless benefited from the workers' weak bargaining power in the labor market and helped perpetuate women workers' subordination (Wright 2001). This labor market vulnerability plays a key role in creating a workforce that is highly dependent on the firm for employment and thus willing to submit to the abuses of a despotic work regime.

Lessons Learned: The Making of the Cavite Export Processing Zone

Although the slack provincial labor market and gendered segmentation made CEZ workers more vulnerable and dependent on their jobs for survival, the coercive labor control at the point of production is in many ways made possible by the wider antilabor strategies of the provincial and local governments, and the reorganization of the zone itself. It is through the actions and participation of a wider-range of actors intervening in a wide variety of domains far removed from the shop floor that we begin to get a clearer picture of the complex and locally embedded labor control regimes that surround the zones (Jonas 1996; Kelly 2001).

Established in 1986, the CEZ is the newest and largest of the four government-run EPZs, with 215 firms employing more than fifty-six thousand workers. The "success" of the zone is due in large part to the lobbying and strong-arm tactics of the former provincial governor, Juanito "Johnny" Remulla (1979–86, 1988–95). After only a year into his first term, the new governor, who had close ties to his fraternity brother, former President Ferdinand Marcos, convinced the president to declare a 275-hectare tract of irrigated rice land around the town of Rosario as the site for the CEZ, which was pitched as a key development to support the central government's export-oriented industrialization strategy. Given the green light from Manila, the governor proceeded to remove and evict reluctant tenant farmers from the prime agricultural land, using bribes and intimidation backed by a private force of armed goons and bulldozers (McAndrew 1994; Dasmarinas 1994). While construction continued, Remulla responded with brutal force in Cavite to the growing national strike wave that shook the Bataan zone and paralyzed Manila. In a strike in the town of Dasmarinas, for example, the governor dispatched local police who waded into the crowded picket lines with batons and gunfire, violently dispersing the picketers while shooting dead one striking worker (Sidel 1999).

The People Power Revolution of 1986 and the rise of President Corazon Aquino initially signaled change in the province as Remulla was removed from office (Sidel 1999). However, Remulla returned to the governorship in 1988 and picked up right where he left off. Although nationally, labor conflicts and general strikes returned after President Aquino liberalized the worst of the antiunion laws, Cavite remained almost entirely unorganized.

Meanwhile, the national government began launching an ambitious industrialization project, funded primarily by the Philippines' largest creditors, the United States and Japan. Remulla again played a central role. The master plan for the new project, drafted from 1988 to 1990 by the Japanese International Cooperation Agency (JICA), made the CEPZ the central axis for Philippine industrialization and growth, thanks primarily to Remulla's lobbying of the Japanese consultants. The plan, know as the CALABARZON project, is heavily biased toward promoting exporting industries, particularly electronics and garments, in the southern Tagalog provinces of CAvite, LAguna, BAtangas, Rizal, and QueZON, just south of Metro Manila (JICA 1990, Ramos 1991).

JICA's master plan set the groundwork for luring more Japanese electronics investments into the country, "promoting the province as a site for Japanese Manufacturers" (FEER 1991, 44, cited in Canlas 1991). The Japanese have been particularly drawn to Cavite because of Governor Remulla, who the Japanese view, according to one writer, as "a political boss who can deliver an inexpensive, compliant labor force" (FEER 1991, 44 in Canlas 1991). It is then no coincidence that Japanese firms have been the leading investors into CEZ. In 1990, for example, the Japanese invested some P548 million, or 60 percent of total investments into Philippine EPZs, primarily into the CEZ (Canlas 1991). By 1999, the Japanese remained the largest investors in the zone (see breakdown of CEZ firms below).

In his second stint as governor, from 1988 to 1995, Governor Remulla was particularly successful in attracting new investors and keeping labor unions out. To help implement his dual strategy, the governor created the Industrial Security Action Group (ISAG), a special police force that also arms and involves local informants and reports directly to the governor. Uniformed police members of ISAG monitor and patrol the various industrial estates and economic zones while local community members of ISAG, according to one official, "act as our peace and order monitors, our listening posts for petty robberies, discordant relations" (Marciano Valando, cited in Coronel 1995, 14). ISAG also provided the muscle to evict a second set of recalcitrant tenant farmers, who were beaten, handcuffed, and dragged from makeshift barricades protecting their lands, during zone expansion in 1991.

At the same time, both militant and moderate unions have stayed away from Cavite during this period, intimidated not only by the ISAG, but also by the "disappearance" and "salvaging" of emergent labor leaders, along with many of the governor's other enemies (Sidel 1998).[2] Little wonder that from 1988 to 1995, only two labor unions were formed in the entire province (Coronel 1995).

The governor also took preemptive measures to enforce his informal but publicly understood "no union, no strike" policy. First, the governor helped arrange job preferences for members of the large and politically powerful religious group, the Inglesia Ni Cristo (INC), to whom the governor maintained close ties. Members of INC are not allowed to join any other organizations, including unions, and are therefore sought as employees in nonunion firms. For example, Storage Ltd. also had an explicit policy favoring members of the INC.

Second, the governor created a standardized bio-data form, which all job applicants must fill out prior to being hired. The form includes employment and background information on the applicant and his/her family members.

---

2. Salvaging is the term used in the Philippines for the murder and public disposal/display of a victim's body, which is often used as an intimidation tactic.

The governor, through his local officials, also made it standard practice that all applicants must have the bio-data form signed by a government office or have a letter of reference from a mayor or the governor. All applicants are also required to have a police clearance, usually obtained from the *barangay* captain, the lowest-level elected official. Workers and factory human resource (HR) managers confirmed that the practice is still standard today.

Finally, the governor also founded and led the Cavite Industrial Peace and Productivity Council (CIPPC), which also included mayors and the heads of provincial investment associations. Ostensibly designed to serve as a liaison between factory owners and the local community, CIPPC in practice acts primarily as an extension of labor regulation for employers. Firms with a reluctant workforce in the face of heavy overtime can go the CIPPC: "We go to the workers and lecture them," stated Maciano Velando, the director. "We tell them, 'This is your company, you should help" (cited in Coronel 1995, 14).

In 1991, Governor Remulla declared the entire province of Cavite an "Industrial Peace and Productivity Zone." By the end of 1995, in his last year as governor, the CEZ was by far the largest EPZ in the country, with 166 firms, employing more than forty thousand workers and did not have a single labor union (PEZA 1995).

Part of the governor's success in keeping unions out is due to where the zone is located, and how is it is constructed and run. Zone planners took particular note of the "disasters" at the Bataan EPZ. First, foreign firms in the 1980s were already moving toward air rather than shipping freight, so the zone was located outside but close to the National Capital Region, which is home to both the Port of Manila and the Manila International Airport. This central location meant a much less isolated enclave and allowed for workers to commute from the surrounding urban and suburban areas.

According to one CEZ administrator, the location in a highly populated area makes recruiting much easier and reduces worker demands on management: "It's easy to get workers, most come from the surrounding area or take the shuttle service [provided by the companies] from Manila. This way, workers can go home to their houses. So even if their salaries are not high, they do not have a lot of expenses." He then compared the situation at CEZ with Bataan, noting, "In Bataan, workers had to come from far away, so the cost of living was high and workers had to spend more and so they demanded more pay."

Second, the zone designers, drawing another lesson from Bataan, purposely avoided constructing any centralized worker housing or dormitories. Again the CEZ administrator noted,

Their main mistake in Bataan was it was too isolated. Everyone had to migrate, which meant they had to create a community *inside* the zone. Unrest at Bataan was high because of the dormitories: it made it easier to organize them and organizers could magnify small problems and recruit new members.

Thus at the CEZ, there is purposely no organized central housing for workers. Workers must find their own housing, usually as "bed spacers" in small makeshift boardinghouses dispersed throughout the local community. These boarding houses, which local residents have hastily added to their existing residences, usually cram six to eight workers into a single room, but nevertheless seldom house more than twenty-five or thirty workers. The boarding house business, along with local food stalls and transportation, are the main contributions of the zone to local residents. While Rosario is a congested, bustling town, workers are interspersed in town and spread across a swath of suburban Cavite. The privatization and wider spread of worker housing thus makes identification of workers and house visits by union organizers more difficult.

Third, unlike at Bataan, the local government officials in Cavite maintain a presence in the zone through representative offices located in or next to the main administrative building. These offices are used to process the necessary town and *barangay* clearances required by all workers and to keep an eye on local constituents. This helps keep a lid on any potential labor problems. The CEZ administrator noted that at Bataan, one small strike erupted into massive labor shutdowns because of sympathy strikes. But at the CEZ, such sympathy strikes are not allowed and national zone authority (PEZA) and the local government work closely together. One PEZA official noted: "If one company has a [labor] problem, the local officials get involved with the solution. The mayor has a lot of say about how these workers behave. If a worker is causing trouble, the mayor talks with the parents or the worker herself."

In 1999, the CEZ was completely full, with more firms and nearly twice the number of workers than any other zone in the Philippines. The Japanese continue to be the leading investors into the zone, owning and or controlling 90 firms. Korean firms are the second largest group, with 70 firms. Filipino firms are a distant third with 22 operations, followed by Taiwanese (20), Singaporean (4), Germans (3), Indian (3), American (2) and Pakistani (1) firms. While initially, the firms represented a balance of garments and electronics, the zone is now heavily tilted toward electronics, representing 80 percent of the locators (PEZA 1999).

When asked why the Philippines, and particularly CEZ, have been so successful in attracting foreign electronics investments, the CEZ zone administrator commented, "It's because we're English-speaking and also the natural flexibility of the Filipino to adapt to any environment." He also attributed the success of the zone to absence of labor problems. As the former manager of the Baguio City Export Processing Zone, which was also hit by the strike waves in the late 1980s, he noted, "We learned to forget about unions and instead put in LMCs [labor management committees]. We push LMCs because they are much better than a full-blown unionized workforce. It's not that I'm antiunion. I'm just pro–industrial peace."

The zone administrator also noted other strategies of labor control at the CEZ. "We have IRD [Industrial Relations Division] that monitors companies and workers for any potential labor problems. PEZA mediates a settlement before it becomes a labor dispute." They also try to intercede before the Department of Labor and Employment (DOLE), which is viewed as more proworker, gets involved. He hastened to add, "But PEZA maintains absolute neutrality between labor and management."

The IRD plays several important roles. First, in order to head off labor problems, the IRD conducts monthly LMC promotion and labor education seminars for new firms. These education seminars are conducted "by nationality," and the IRD hires interpreters to explain, in the managers' language, the Philippine labor code and labor laws and well as explain the "culture" of Filipino workers. The IRD chief, in fact, noted that cultural differences in handling employee discipline were often at the root of many labor complaints, adding in a hushed but irritated tone, "The Koreans are the worst." This anti-Korean prejudice, often based on rumors, was nevertheless prevalent in many interviews.

The IRD is also central to employee screening. Applicants who flock to the zone in search of work are not only required to have local government and police clearance, but are also coursed through the IRD office for endorsement to firms. This "manpower pooling" is a major activity of the IRD, which tries to "match the skills" needed by firms with those offered by job seekers. The pooling also allows PEZA to keep track of workers in the zone. The chief of the IRD complained that she has more than one hundred applicants a day coming to her office in search of work.

When asked what kind of workers "match the skills" needed by firms, she explained that firms are looking for women, because they are "industrious and patient," between eighteen and twenty-two years of age, because the "young ones have better eyes and dexterity," and at least high school graduates. Both the zone administrator and the IRD chief noted that while garment producers often are willing to hire older workers, "even twenty-five-year-olds" with sewing experience, the electronics companies' main hiring criteria is "trainability." In fact, the population of zone workers reflect these criteria: 88 percent of those employed in the zone are factory workers, 73 percent of all workers are female, 71 percent of workers finished high school, while nearly 17 percent had either some college or has finished a college degree. On the other hand, more than 95 percent of the more than six hundred foreign management personnel are male (PEZA 1999).

A more in-depth survey of 333 workers from 55 firms at the CEZ conducted in 1996 found that only 20 percent of the workers came from the province of Cavite and only 5 percent from the town of Rosario. The vast majority were migrants from other provinces. The study also found that 65 percent of the respondents were between nineteen and twenty-five years of age, while 14 percent were between sixteen and eighteen. Only 21 percent of the workers were older than twenty-five. For 54 percent of the workers, their present jobs were their first experience in the labor force. In terms of pay, more than 38 percent of the respondents reported that they were being paid below the minimum wage of 155 pesos per day, 40 percent said they received the minimum, and only 23 percent stated they were receiving above the minimum wage (WAC 1996).

The strong-arm industrialization strategy begun by former governor Remulla has proven quite successful in drawing large numbers of foreign electronics firms into the province and into the CEZ. However, despite the complexity of formal and informal labor regulation and sometimes brutal working conditions within the zone, the system is not without conflict, and conditions have proven to be neither entirely stable nor static, as political changes have shifted the institutional and regulatory landscape in Cavite. First in 1992, although Governor Remulla easily won reelection, he backed a losing national presidential candidate, thus failing to build patronage ties with the new Philippine executive, Fidel Ramos. Ramos, trying to put his own party in power, moved to oust Remulla and backed Epimaco Velasco in the 1995 governor's race. Velasco, with his own provincial base and campaigning on a proworker platform to build support from the burgeoning number of workers, succeeded in toppling Remulla (Aganon et al. 1997). But for the new governor, getting things done in the province was difficult, since the vast majority of mayors and *barangay* captains still belonged to Remulla's Magdalo political party. His proworker campaign promises also proved to be mainly rhetorical. As the head of a church-based labor organization explained: "Remulla was tough, he started the 'no union, no strike' policy. Velasco, he had the ear [for listening to prolabor groups] but not the muscle. With him, the policy became 'unions, but no strikes.'"

Nevertheless, the 1995 ouster of Governor Remulla led to an incrementally wider political space for labor that has reverberated down to the local level. Emboldened by then-governor Velasco's prolabor rhetoric (if not his actions), both moderate and militant unions reemerged in the province (ICFTU 2003; WAC 2003). Their active organizing and links to zone workers have led to an increase in collective labor activism within the zone, including the zone's first strike by the workers at Power Tech in 1997, discussed in the previous chapter. By late 1999 eleven unions had officially registered with the Department of Labor and Employment (DOLE). Only three, however, had won certification elections and none had yet negotiated a collective bargaining agreement (CBA). One of these unionization drives, at Allied Ltd., will be detailed in chapter 5.

As organizing continued in the zone, so did workers' strikes, disruptions, and management reprisals. In 2000, the zone experienced at least four strikes, in both electronics and garments factories, leading the firms, the provincial government, and PEZA to step up their labor control and anti-union strategies. For example, one of the most effective tactics is for a firm to simply suspend operations during union formation, then resume production at a "new" firm under a different name. As one exasperated labor organizer stated, "This is their most effective union-busting strategy. They just shut down the factory and change the corporate name. It's the same product, same workers, same management. Only no union."

This is exactly the tactic that Power Tech management employed. After the company transferred a portion of production and a large number of union member to the nonunion Ultra Electronics, the former Power Tech workers began organizing again, and succeeded in establishing a new union at Ultra in May 1999. Power Tech, as a majority stockholder, then dispatched its HR manager as an "HR consultant." Management promptly accused the union president of theft and suspended him. They then filed for a temporary shutdown and stayed closed for four months. When they did reopen, they immediately retrenched ninety-nine remaining union officials and members and went into slowdown mode and eventually closed again. And in what is now a clear pattern of behavior, a major Korean shareholder from Ultra with close ties to the Power Tech group set up a new, nonunion firm also in the CEZ that is beginning to produce the same products as Ultra did before its shutdown (WAC 2001).

As firms came up with new responses to labor organizing, major changes also took place in both zone administration and in provincial politics. First, PEZA officials began publicly blaming the slump in zone investments in early 2002 on the renewed "labor activism" in the zones that was supposedly scaring off investments. In the zones themselves, PEZA officials has already reversed their former policy of allowing unrestricted public access into the zone, instituting a new zone pass for anyone coming into the zone. To get a zone pass everyone has to register with the zone administration, submitting a complete bio-data form, including fingerprints, identifying marks, photos, community affiliations, a telephone number, and a local address. Organizers claim that the pass is used primarily to recognize and bar union organizers and to intimidate workers. Zone officials also centralized all hiring, requiring firms to post openings in a central, administrative area where initial screening/interviews are also conducted.

However, changes at the provincial level have had the strongest (negative) impact on labor organizing. In 2001, gubernatorial elections in Cavite Province proved once again decisive. A congressman and close associate of former governor Remulla, Ireneo "Ayong" Maliksi, ran explicitly on a labor control platform. It is worth quoting at length a letter, written on official stationary, that the congressman sent to the president of the Cavite Export Zone Investors Association dated March 26, 2001, soliciting the support of the association of investors.

> I write to inform you that I have filed my certificate of candidacy as a standard-bearer of former Governor Remulla's PRATIDO MAGDALO, and is [sic] seeking the gubernatorial post for the Province of Cavite.
>
> I seek this post with great credentials from the history of our party, which spurred the industrialization of the Province. It is sad to note that in the past 6 years of Cavite's leadership under a different party, it seems that the industrial sector has been set aside. It is specifically painful for us, considering the efforts we have undertaken so that industries like yours could be established and protected in Cavite. This is why it is on the top of my list to reinforce the industrial sector, and restore industrial peace as was under the MAGDALO leadership. Through the "No union No-strike" policy which our party has succeeded in implementing up to 1995, we shall be assured that no disruption of work shall occur, and that profitability of your operations are increased. Labor groups such as [x] shall no longer be a problem since majority of them are endorsing my candidacy. . . .
>
> I believe that we in the Partido MAGDALO can improve the situation of industries far greater than my incumbent opponent. We have experience, we have the know-how and we have the will.
>
> This is why I am seeking your support in the upcoming election. Together, let us restore Cavite's industries to its rightful place.

Maliksi went on to win the election and indeed vigorously pursued an antiunion campaign. Like former governor Remulla, Maliksi set up an intelligence and policing network, this time under the new name of the Cavite Industrial Peace Advisory Group (CIPAG). CIPAG tapped into local police and mayors' networks to collect data on union leaders, while also employing a cadre of ex-union organizers to promote the formation of labor management committees (LMCs) at firms experiencing a union organizing drive. CIPAG members were accused of directly intimidating workers during a union drive at an electronics factory in one of the province's many private EPZ. Armed with a company-supplied list of active unionists, a team of CIPAG members allegedly conducted house-to-house visits of union members and leaders, threatening that they would lose their jobs if they did not sign an affidavit renouncing their union membership (WAC 2002). As one union organizer summed up the situation, "It's like going back to the days of Remulla. But now it's even more comprehensive. PEZA couldn't handle all the action themselves, so they called in the governor's office."

Despotic Control, Low Commitment, and Instability

The recent problems at Power Tech and the rise in union organizing at the CEZ in general highlight the potential instability of both shop floor despotism and coercive external labor control. As Jonas (1996, 228–29) points out, the regulation of labor and the labor market, what he calls a local labor control regime,

> is not a static and fixed object but a rather fluid and dynamic set of social relations and power structures which are continuously reproduced and/or transformed by forces of domination, control, repression and resistance operating at a variety of scales. In other words, local labor control regimes are negotiated and contested socio-territorial structures.

Indeed, the labor control strategies at Allied-Power, while multilayered and stretched far beyond the shop floor, are not necessarily stable. The rising labor conflicts in the zone clearly demonstrate that even sophisticated labor control regimes are contingent and subject to challenge. Just as the Bataan zone was originally designed to contain labor conflicts, the CEZ too was considered a new, improved model of zone organization that could deliver on its promise of "no unions, no strikes." But the CEZ, as a public zone, is still formally accountable to local citizens, and has the explicit mandate to provide employment and foster local development, a mandate it shares with the other three public EPZs. This mandate has helped create the conditions for potential problems, since to fulfill it, the administration has sought to attract a large number of labor-intensive firms. Meanwhile, because it is a government-run zone that is publicly answerable to a shifting political base, the government authority that manages the zone is now somewhat limited in their labor control strategies. Thus the high concentration of labor-intensive firms in a public, government-run zone has likely contributed to both the rise in conflict and an inability to entirely contain it.

From Sweatshop Platform to High-Tech Enclave: Storage Ltd. and EPZ Privatization

> Leaving at dawn today, I traveled over two and a half hours from the Storage Ltd. factory complex on a recruiting trip, over highways, along country roads and into a far corner of the rural province of Quezon. The company and the local mayor's office set up the Job Fair, one of three for this week. Two other companies are recruiting: one looking for machine operators at a biscuit plant and the other for maids and factory workers to do overseas contract work in Taiwan. By 8:00 a.m., the town's shaded basketball court was packed with hopefuls and on-lookers. First, the recruiter di-

vided the crowd of maybe 100 at half court; one side men, the other side women. She explained they were only recruiting women today and that the men—who made up over half the group—could leave. She then collected the women's bio-data sheets they had all pre-pared. The women, mostly daughters of farmers and housekeepers, sat nervously on folding metal chairs, smiling and hopeful. After scanning their applications, the recruiter quickly reduced the large pool to only fourteen because most did not fit the narrow age slot (eighteen to twenty-two), or meet the education and (in)experience criteria. These applicants were then given a battery of tests to evalu-ate their math, English, nonverbal reasoning, and speed and accu-racy skills. Eight were selected to sit for five-minute interviews.

In the interviews, the recruiter asked in rapid-fire succession about the ages and jobs of the applicants' parents and siblings, willingness to work overtime, willingness to relocate, and asked, "Do you know what a production operator does? Is screwing in screws OK with you?" They were also quizzed about their attitudes towards unions: "What do you think about unions? Would you ever join a strike?" All but one applicant was hired. When I asked her later why the one applicant was rejected, the recruiter replied that she appeared too nervous and had worked in Manila before. In addition, the recruiter noted, "She said she was 'OK with a union' and she looked too skinny to render overtime."

Field notes, 16 July 1999, Quezon

In contrast to Allied-Power, Storage Ltd.—the hard disk drive producer—has a much more expensive and sophisticated production process and has chosen a panoptic work regime that combines marginalizing production uncertainty with "disciplinary management." Yet the firm also tries to minimize the wages and costs associated with its more than nine thousand employees. Such a tightly controlled and low-paying labor control regime would seem incom-patible with flexible high-tech production, which demands some level of worker commitment for stability. So how does the company accomplish this?

Management at Storage Ltd. has essentially "solved" the incompatibility dilemma by extending labor control from the production locale (the factory) to wider institutional domains of labor control, namely reproduction locales such as the local labor market and local governments, and consumption lo-cales such as communities, families, and boarding houses (Jonas 1996). Specifically, the firm taps into (and helps construct) local labor market seg-mentations, while also benefiting from the new privatized and spatially re-structured EPZs, selective state intervention, and links to other formal and informal local institutions. By exploiting workers' weak bargaining power both inside and outside the plant, the firm is able to secure a kind of alien-ative or "coerced commitment."

PEZA, Private Zone Developers, and a New Regulatory Model

A fundamental difference between Allied-Power and Storage Ltd.'s localization strategies is related to the nature of the zones in which they operate. Storage Ltd., like Integrated Production and Discrete Manufacturing, chose to locate in one of the Philippine's forty-seven active special economic zones (SEZs), which are all privately owned and managed, mainly by Filipino real estate developers (PEZA 2004).[3] This is in contrast to the four older government-operated zones in Baguio City, Bataan, Mactan, and Cavite. The privatization of the zones was central to the national government's revised development strategy in the mid 1990s, which aimed to swiftly liberalize the economy and privatize many of the government-run businesses (Hutchcroft 1998). The government policy switch from building large public zones to promoting smaller private zones also reflects the changing needs of investors, and the Philippines' attempts to remain competitive among other investment-hungry countries. In 1995, the national zone authority was reorganized and renamed the Philippine Economic Zone Authority (PEZA).

As part of its new development strategy, PEZA immediately targeted larger, more technology-intensive manufacturing to anchor its revamped zones. However, under increasing neoliberalization and pressure to privatize, the agency did not have the resources nor the expertise to build the state-of-the-art infrastructure, such as specialized power and wastewater treatment plants, or digital communications systems that are necessary for advanced manufacturing. As an official of national zone authority remarked, by the mid 1990s the government had limited resources and could not meet the demands of the bigger manufacturers it hoped to draw. In discussing the older zones and the need for change, the official noted,

> Our budget [for the four public zones] was all loans from Japan—JICA loans [Japanese International Cooperation Agency]—and we're still trying to pay for those zones. So the burden was on the Philippine government and the Philippine people. Those zones were like white elephants, and by the nineties, you could see the decay. . . . The innovation came when the government opened the door to private participation to develop the new zones. . . . We wanted "world-class" zones and public zones are just too bureaucratic.

The privatization and decentralization of the zones also meant the entrance of new actors and a shift in how the zones are organized and regulated. While the government was quick to create the laws for privatized zones, Philippine law still prohibits 100 percent foreign ownership of land.

---

3. As of September 2004, the Philippine Economic Zone Authority managed a total of sixty-five operating zones, four of which are publicly owned and fourteen of which are actually information technology (IT) buildings, which host no manufacturing and instead cater to offshore IT services such as call centers (PEZA 2004).

Thus the major players in the newly created and high-stakes market for private zone development have mainly been large Filipino real estate companies. Established Filipino land and property developers such as Geronimo de los Reyes or Jamie Zobel de Ayala sought partnerships with foreign investors/developers who could provide the capital, expertise in infrastructure development, and connection to foreign manufacturers (Tiglao 1996). Many of these foreign partner firms have been Japanese and Singaporean firms with previous experience in zone development elsewhere and/or marketing connections to potential locators or manufacturers. Since the mid 1990s, Philippine developers have gained enough experience to plan and develop the zones on their own, yet they still need foreign partners to help finance the zones and for marketing abroad. Prompted by the government's privatization scheme, local developers have also banded together and created their own organization, the Philippine Ecozone Association (PHILEA), with twenty-nine members, twenty-eight of whom run PEZA zones. The association coordinates many developer activities and maintains an office at both the PEZA national headquarters and the government-run Board of Investments office.

By 2000 there were officially 130 planned and operating special zones spread throughout the Philippines. But by 2004 only 35 percent of these were fully operational, and the bulk of the active zones are still found in Region IV, know as the Southern Tagalog region. The concentration is mainly due to the integration of private zone development into the state's official CALABARZON development project, discussed above, and the proximity to the central airport in Metro Manila.

The private local zone developers and their foreign partners work closely with PEZA. As a representative of PHILEA noted, "We're the private arm of PEZA . . . we have a very close working relationship, but we maintain proper distance." This is particularly clear in the marketing of their zones—and the Philippines generally—to foreign investors. Local developers and representatives of PHILEA often accompany the director general of PEZA on "locator missions" abroad, and commonly join officials from the Department of Trade and Industry and/or the Philippine president on state visits and "road shows" around the world. When asked what foreign locators are looking for and how local developers market their zones, a PHILEA representative responded that stable macroconditions were vital: "infrastructure, like power and roads, and peace and order." In terms of labor, he ticked off the results of an American Chamber of Commerce study ranking investors' highest concerns: "security, peace and order, and [lack of] corruption. . . . That's their priority, so it's ours too . . . labor costs are fourth or fifth. Gone are the days when we could brag about our cheap labor. We can always brag about our skilled labor, but our neighbors are catching up."

The changing slate of pertinent actors has also led to the reorganization—

but not the deregulation—of the zones themselves. For example, one of the major differences between the public and private zones is its administration. While the public zones are entirely run by government officials of PEZA, in private zones there is only one PEZA zone administrator with a small staff that mainly facilitates customs clearance. As a central PEZA official expressed, "The government doesn't meddle in the private zones. The zone managers only handle day-to-day matters." Indeed, the new private zones have reduced the amount of direct government oversight by individual federal government agencies that are active in public zones, such as the Commission on Audit and the Department of Labor and Employment. Instead, a single government zone administrator and a small staff act as a "one-stop shop" for firms dealing with all government bureaucracies. The PEZA administrator at the private zone where Storage Ltd. is located boasted,

> We [the government zone authority] have a commitment to locators that they deal only with one government agency, so if there is a problem, like a minimum wage problem, we deal with it first. We have an understanding with the Department of Labor and Employment and their inspectors; they can't come in here without going through us first.

Privatization has thus led to *re*regulation, which allows the new private zones to step away from the explicit employment and local development mandates of the public zones, as well as insulate themselves from direct, public scrutiny and accountability for conditions within the zone.

In its privatization strategy, PEZA was keen to avoid the "mistakes" made at both Bataan and Cavite, particularly in terms of spatial zone design and layout. First, PEZA wanted to prevent the high density of firms and overconcentration of workers, such as found around the town of Rosario and the Cavite Economic Zone. PEZA thus encouraged the real estate developers to construct a large number of zones, each with relatively few locators, scattered in less densely populated areas. For example, whereas the Cavite Economic Zone has 215 firms and employed 56,000 workers, the private zone where Storage Ltd. is located has only 17 firms and a total of 16,000 workers.

The new zones have also tried to improve on the worker housing strategy at the Cavite Economic Zone. There, to avoid the centralized worker housing model that ended up helping union organizers at the Bataan zone, PEZA provided no centralized housing at all. Yet this deregulated housing policy led to the unplanned concentration of workers in the communities immediately surrounding the zone. At the new private zones, in contrast, PEZA has actively *dis*organized worker housing by placing the zones further away from residential areas. Instead, both PEZA and the zone developers encouraged firms to provide shuttle bus service from multiple pickup points located near existing but widely dispersed population centers. Wide, provincial recruitment coupled with the firm's free shuttle bus service means that the workforce has less in common and is less concentrated in any one area. Storage Ltd., for example, deploys some seventy-five buses to thirteen different daily pickup points spread out in all directions up to two hours away. Workers are encouraged to commute to or live in a boarding house near one of these thirteen pickup points—greatly dispersing the workforce and making union organizing extremely difficult. This housing strategy, devised by the national zone authority, is now commonly used in the other private economic zones.

Once locators invest, the zone developers continue to play a key role for both the foreign manufacturers and the Philippine state, acting as de-facto regulators and local "service companies." The zone developers provide three main services: infrastructure, security and labor recruitment, and community relations. One of the most important "selling points" PEZA and developers can offer to foreign investors is the promise of a "strike-free" zone. Given the specter of the strikes and unionization drives in the Bataan zone, nearly all foreign electronics investors fear unionization in the Philippines. As noted in the previous chapter, Japanese investors, who now make up the largest share of foreign investors in the electronics sector, have made it clear to officials that they will not locate in the Philippines without a virtual guarantee of union- and strike-free zones. Since the zones are privatized, the developers become involved in and even responsible for ensuring "industrial peace."

Securing "industrial peace" around the zones involves an array of actors and a host of strategies. First, security in the zones themselves is very tight. Whereas the Cavite Economic Zone is publicly accessible to anyone, including job seekers, private zones allow no such "walk-in" access. In the zone where Storage Ltd. is located, there are two armed check-in and inspection points where visitors must surrender valid identification cards even before reaching Storage Ltd.'s own fully secure area. No one is allowed into the zone if they are not a direct employee or have specific business with a locator. Vehicles are physically inspected on both entry and exit. Known labor organizers are barred from entering the zone at all.

Labor and union control strategies around the private zones are in many ways similar to those at the CEZ. However, PEZA, PHILEA, and other government actors have also developed much more sophisticated strategies beyond simple coercion or strike-breaking. An official of PHILEA noted proudly, "Strikes are down to a bare minimum" in the areas around the zones. "People now are careful. They understand that it's better to have work than no work, better to have less pay than no pay." But in addition to highlighting the threat of firm exit and slackness in the labor market, PHILEA and PEZA also pursued a "proactive strategy" to discipline workers.

A PEZA attorney in the labor relations department also highlighted the drop in strikes around the newer zones, but attributed the decline to several new initiatives of labor education and cooperation with local officials that recognized the importance of "labor relations beyond the workplace and workplace issues." This attorney noted that workers may not get all they need at the workplace, and "so now, the community gets involved in lean-periods." Specifically, PEZA along with the local government units (LGUs), which include town mayors and *barangay* captains (the lowest elected officials, similar to "village heads"), promote the creation of labor management committees (LMCs) that extend beyond "labor" and "management." In the province of Laguna, where Storage Ltd. is located, the provincial government has helped create LMCs that include representatives of workers, enterprises, LGUs, PEZA, the Department of Labor and Employment, and zone developers. As the PEZA official from the labor division noted, "This way, the enterprise can count on the LMC and the government to resolve disputes and ensure it won't ripen into a full-blown strike."

Missing from the PEZA representative's account was the crucial role the coercive arm of the state plays in keeping the "industrial peace" around the zones. The active role of the local and provincial police force was outlined more clearly by a representative of PHILEA. He noted that in Laguna, where Storage Ltd. is located, the group participates on a council with members of the Philippine National Police (PNP). "We [the zone developers] present our concerns and they brief us on the security conditions for a more proactive strategy on security." He also explained that the zone developers cooperate at the provincial level with the PNP in order to "make sure unions in the parks [zones], if there are any, are not infiltrated by left-leaning groups. We're building up employee profiles through PNP to keep tabs on people."

Labor control and antiunion activity is closely linked with the other major role of the private zone developers: helping develop relations among locators within the zone and with local government officials for labor recruiting. The zone in which Storage Ltd. is located is typical. The developer there has organized the only zonewide group of locators, the Human Resource Association, which brings together human resource managers, the government zone administrator, and the zone developer for monthly meetings. Participants exchange information about available workers and recruitment strategies, but also deal with potential labor problems and circulate a black list of "troublemakers" and union sympathizers. The developer then helps the HR association make contact and maintain relations with government officials at the provincial and local levels to help source manpower.

Links with local officials are crucial, particularly for recruiting. Job fairs are set up in the provinces with help of local *barangay* captains, mayors, and the Provincial Employment Service Office (PESO), whose staff locally is appointed by the mayor. Mayors and *barangay* captains are keen participants in recruiting, because as elected officials they want to be viewed as the link that got locals their jobs. Like firms in the Cavite Economic Zone, Storage Ltd. requires a letter of reference from either the mayor or *barangay* captain, and a local police clearance. Although the mayor or captain may not be directly responsible for getting locals their jobs, they are necessary intermediaries for workers and one more figure to which workers owe a personal debt, usually paid back in votes. One mayor's information officer noted that they receive up to fifty recommendation requests a day and keep a complete and active database on all those for whom the mayor has written letters. In return, the company knows it can rely on the officials if they have any labor organizing problems. The human resource manager at Storage Ltd. boasted, "We work with the *barangay* captains because they practically guarantee their recommendations and whatever happens, they are responsible. They certify that the person is of good moral character and will always be there to help if there is any union activity."

This strategy was echoed by the PHILEA representative, who stated that working with mayors and *barangay* captains was, on the one hand "good— it gives us access to manpower and if the guy goes bad, we can go back to the mayor and say, 'Hey, this is the guy you recommended and now he's the one forming a union!" But on the other hand, working with LGUs "can also be bad. There are so many people and only few jobs, we have to say no sometimes." Indeed, for the zone developers and firms themselves, working with mayors and local officials can also be problematic, since mayors often push to secure both direct and indirect job allotments to boost their patronage. In terms of direct employment, they try to get preferences for actual production jobs in the firms. In many large, multinational firms that have rigorous screening, this is not often forthcoming. But a much more lucrative business for local officials are the indirect employment opportunities; such as getting contracts for zone security, catering, or janitorial services. As the PHILEA representative stated, "The LGUs come to us [zone developers], and we do have moral persuasion with locators. For indirect jobs we say to [locators], 'Get it from the mayors.' But it's their jobs and their decisions."

The Outside Game: Reaching Beyond the Plant

Just as the heavy-handed labor control strategies of the local government in Cavite help make possible the despotic work regime at Allied-Power, so too does the more sophisticated system of labor control surrounding the privatized zones provide support for the external strategies of Storage Ltd.'s panoptic work regime. As was made clear in the last chapter, Storage Ltd. has developed a complex system of shop floor and internal plant discipline based on visibility, surveillance, and fear. Yet internal strategies by no means exhaust Storage Ltd.'s repertoire of control and commitment strategies. In fact, the company's most effective strategies for maintaining their union-free status and a very low turnover rate of only 5 percent a year is related to their gendered recruiting and external labor relations practices.[4] Like Allied-Power, Storage Ltd. recognizes its deep social and political embeddedness and takes active steps to localize its operations to take advantage of location-specific conditions. However, the character of Storage Ltd.'s embeddedness is somewhat different than Allied-Power, due in part to the nature of its production, the intensity of its HR strategies, and the nature of the economic zone and its environs.

RECRUITMENT   For recruiting new operators, Storage Ltd. does not rely on an outside hiring agency, like some firms, but hires all of its operators directly. And unlike Allied-Power, which relied mainly on walk-ins or word-of-mouth, Storage Ltd. put a great deal of effort and resources into actively recruiting and screening its potential employees along detailed criteria and from a wide, geographic area. During the time of field research, Storage Ltd. was expanding rapidly, hiring at times 200 to 250 new operators a week.

Storage Ltd.'s most basic hiring criteria are gender, age, and education level. On a large banner hanging from the side of their main factory building, their criteria are announced quite explicitly:

Now hiring production operators:
- female
- 18 to 22 years old
- High School graduate (some college or vocational school preferred)
- No experience necessary
- 5 feet tall

In fact, the profile of the company's nearly nine thousand workers is remarkably uniform: 88 percent are female, 83 percent are between sixteen and twenty-two years old (29 percent between sixteen and nineteen) with an average age of twenty-one. Fifty-one percent have some college or vocational training (the other 49 percent are high school graduates). One of the company's other main criteria that is not posted in the banner is marital status: 97 percent of the workers are single.

The reasons management gives for this rigid hiring profile are generally the same as those at nearly all other electronics firms in the Philippines and, indeed, in zones around the world (Ong 1987; Chant and McIlwaine 1995; Hossfeld 1995; Lin 1997; Yun 1995; Lee 1998; Chhachhi 1999). Manage-

---

4. Overall, Storage Ltd.'s turnover rate was 14 percent in 1999, but this included probationary employees who had been with the firm less than six months. Among regular or permanent employees, the quit rates dropped dramatically and the turnover rate here is the quoted 5 percent. Permanent employees make up 70 percent of the workforce.

ment says it prefers women because the work requires "patience, dexterity and attention to detail." Men are viewed as lazy, sloppy, fidgety, error prone, and generally more *magulo* or disruptive, but the connotation here is that they can be troublemakers.

They prefer single women because, as one recruiter put it, "if they are married, they need time for kids and husbands." Workers interviewed also clearly understood this hiring criteria and the reasons behind it. Graciela, an unmarried female technician: "For operators and technicians, they don't hire anyone who is married. If you're married, you would not really be committed to work anymore, since you'd have a family to look after." This was echoed by Irma, a single production operator in slider assembly: "No, they don't take married women, they accept only those eighteen to twenty-two and single. If you're married you'd have family responsibility so you'd have to split your time." In fact, every operator and HR staff person interviewed reaffirmed this (quite illegal) hiring practice.

In terms of education level, the company wants "balanced" operators who are intelligent and English-speaking, but not "overly educated." All operators must also go through three months of technical training and almost all instruction materials are in English. At the same time, the firm does not want operators with high job, pay, and promotion expectations. For this reason, at the operator level, they do not hire anyone with more than two years of college. Any applicant who is a college graduate will only be considered for a position as a technician or engineer's assistant. And the preference for these positions is typically for men. As pointed out in the previous chapter, this strict "window" of educational requirements effectively segments the workforce by gender and severely limits operators' mobility.

The experience criterion is linked to that of age and education. In fact, having "no experience" is not only acceptable it is preferred. The key is "trainability." Storage Ltd., like many other firms, prefers "fresh workers," that is, those with no more than one year with one employer and no more than two years' total work experience. The reasons are twofold: First, as the recruitment manager stated: "We want to be the first one to mold them. They are easier to influence according to our culture, and its tough to break old habits formed in another company." An Employee Relations staff person added: "In hiring experienced workers, the main problem isn't about training but about values and attitudes. We are trying to protect the culture of our company. Fresh workers are better for organization and harmony between employees and management." But the HR director was more forthcoming. He stated simply, "We want to eliminate those who might have union affiliations."

Avoiding potential union sympathizers is another "hidden" but quite clear hiring criterion. Here, there are three main strategies. First, the firm avoids hiring anyone who has worked for a company known to have militant union, whether or not the applicant was a union member. Second, the firm gives preference to members of the Inglesia Ni Cristo (INC) church. Recall from the section above, this was a practice used by former Cavite governor Remulla since church members cannot belong to other organizations, including unions. Storage Ltd. employs more than five hundred members of the Inglesia Ni Cristo. Finally, in the interview, the firm asks questions to assess an applicant's proneness to joining a union. One interviewer told me her technique: "I always ask it this way; 'It's OK for you to join a union, right?' because people always say yes to a recruiter. This way I test them. If they don't answer, I then explain more about unions: 'Would you join a group that goes on strike to raise its wages?' Most then know what is being asked and say no." Another technique used is for the recruiter to ask, "What do you think are the rights of workers?" If the applicant knows too many rights or answers aggressively in favor of unions, they are not hired.

Workers confirmed that the company was quite explicit in its antiunionism. Operators interviewed all stated that during their interviews, they were asked what they thought about unions and all understood that a "wrong" answer would mean they would not get the job. Kiko, a warehouse operator whose girlfriend also worked at Storage Ltd. and helped get him his job said: "When I was interviewed they asked me [if I though unions were okay]. I answered, it depends. I saw they reacted to that, so I took it back. But they didn't turn me down because I had a backer." Others simply replied that unions were "forbidden" (*bawal*).

The interview is also used to assess "attitudes and values" and focuses on an applicant's family and role in the household. In a very conversational tone, the applicant is asked about her family background, the age of each family member and what each does. While seemingly innocent (talking about one's family in the Philippines is akin to Americans talking about the weather), the line of questioning reflects another of Storage Ltd.'s hiring preferences. When possible, Storage Ltd. looks to hire the oldest sibling in a family, because the eldest are usually responsible for helping parents financially and helping pay the tuition and school fees of their younger brothers and sisters. They also check if the parents are working. It is a plus if the parents are unemployed, because as one staff member commented, "breadwinners stay longer. And besides, hiring the oldest or someone who will be the sole family breadwinner is like charity work." Thus the goal is to see how dependent the family is on the wage of the Storage Ltd. worker and how family life might interfere with work.

To recruit workers, Storage Ltd. relies mainly on two sources: referrals and job fairs. Current workers are encouraged to tell their family members about openings and receive P100 for each person they refer that is hired. About one third of new hires come from worker referrals. Referrals are used as a screening and disciplinary mechanism, since workers only refer those who will not make them look bad and new hires will have *"utang na loob"* or personal debt to the current employee who helped get them a job.

Job fairs help recruit almost half of all new hires. Recruiting is done mainly in the northern and southern Tagalog-speaking provinces just to the north and south of Metro Manila. In a two-month period, from June to July 1999, recruiters made more than fourteen recruiting trips to provinces as far away as Palawan Island, Bulacaan, Pangasinan, Quezon, and Batangas. Other trips were made to rural areas in the nearby province of Cavite and within Laguna itself. Storage Ltd. used to recruit from the Visayas, a large group of islands located in the southern Philippines, but stopped because of workers from those areas spoke a separate language, which caused communication problems with supervisors and fear that those from the Visayas would form a tight, self-identifying group within the plant. Still, the firm favors hiring from the provinces because they believed that there are fewer opportunities in rural areas. Workers from the province are seen as "the shy type with no qualms about following orders. They work harder, are more focused, and work more seriously."

The matter of fewer opportunities was well born out when I accompanied recruiters to the provinces. The local officials were often excited, because very few other recruiters were hiring for positions that lead to permanent jobs. Most other manufacturers, they said, are looking only for contractual employees. But local officials were concerned about the age and gender requirements, since most locals looking for work were male or older females that were retrenched elsewhere and were now "too old" at twenty-three to get another job with a semiconductor firm.

Barangay captains are the most important and most useful community contact for the company because they are less busy than mayors and know their constituents better. But relations with the *barangay* captains are not always harmonious. One HR staff commented, "There are many political reasons to coordinate with local *barangay* captains. If we don't help them, they can make things difficult." Pressed for details, she explained that the developer promised the relocated tenants, who used to work the land that is now the EPZ, that their kids would get jobs in the zone. Some *barangay* captains came directly to the office of the HR manager requesting jobs and a few actually threatened the company with sabotage during the construction phase. To diffuse the tension, the company now gives informal priority to locals around the zone and holds job fairs there. One solution was to hire the bulk of the janitors and drivers from the local *barangay*. However, these jobs were subcontracted through a hiring agency and thus were contractual and not permanent jobs.

BACKGROUND INVESTIGATION   Recruiting is linked very closely to Storage Ltd.'s Background Investigation (BI) Unit, an important extension of the firm's panoptic work regime. Again, unlike Allied-Power and the other firms in this book, Storage Ltd. went to great lengths to collect detailed information about its workers and their lives far away from the point of production. As mentioned before, Storage Ltd. hires no contractuals. But, by Philippine law, anyone who is working fulltime and doing the work of a permanent employee must be made a permanent employee after six months. Thus for the first six months workers at Storage Ltd. are probationary. Once a worker becomes permanent, they get full legal protection under the Labor Code and are more difficult to fire without just cause. Consequently, before workers become regulars, the Background Investigation Unit is called in to verify employee data, gather more detailed information about workers, and make an appearance in the workers' parents' home and community. Primary targets for investigations are those employees who have been reported as being under- or over-age, or as having married since being hired. When asked about the marriage policy, the BI staff person remarked, "Employees are not allowed to get married during their probationary period because, you know, married women get pregnant."

Referring back to bio-data forms originally submitted during hiring, the background investigator sends out letters to check a worker's birth certificate, character references, schools records, *barangay* connections, and family situation. They check specifically the worker's real age and civil status, and follow up on workers who are making benefits claims, like hospitalization, sick leave, or maternity leave.

If the background investigator finds inconsistencies, it is grounds for dismissal. Though patently illegal, the company enforces strict hiring criteria based on age, sex, and civil status. Probationary employees have little legal recourse and realize the very vulnerable position they are in and thus understandably might withhold any information that would jeopardize their current employment and the possibility of becoming a regular, permanent employee. By making their (illegal) hiring criteria explicit but making the falsification of employee records as the offense punishable by termination, the company can avoid possible discrimination and illegal termination suits. Thus a probationary employee cannot be formally fired for being married or being pregnant, but if she does not report her change in status, she can be terminated for "falsification of company personnel records." If she does report it while still a probationary employee, she has few legal rights and can be easily terminated, even without "just cause."

Such a case was documented in a BI report to the HR director in December 1998:

Statement of Problem: Due to the increasing number of pregnant employees, the undersigned checked the records of employees who availed of the SSS [Social Security Service] maternity benefits. As a result, it revealed that [name of employee] was one of the married employees who did not declare her true civil status upon employment with Storage Ltd.[5] In this case, the employee was confronted and "asked to resign."

Probably the most common "discrepancy" for background investigators is underaged workers falsifying their birth certificates. The following is from a BI report, dated April 1998:

> BI Findings: An anonymous tip named [employee] as underage and when interviewed, [employee] admitted to lying about her age. [Employee] explained that she needed the job very badly because during that time, their house was recently destroyed by a fire. She furthermore explained that she is the eldest among six children of [name of employee's father] who is half paralyzed by stroke and [name of mother], who is working only as a farmer. Three of her siblings are still in elementary school.
>
> Recommendation: Before this report was made, [name of supervisor] called up the undersigned [HR assistant] to ask if we can give some consideration to [employee]'s performance, since she is an excellent operator (see attached letters from [employee]'s line leaders); in addition to her reasons for committing such violations. Nevertheless, subject employee is still recommended for termination for violation of company policy of "falsification of company personnel records."

Asked about this particular case, the background investigator who wrote the report defended her actions: "We get a lot of cases like that. They are the oldest, their parents are farmers, they have six brothers and sisters. It's tough. Sometimes I cry. But we have to do it. It would be unfair to others if we didn't fire them. . . . That's why they call us 'the terminators.'"

The main tool of the background investigators is the home visit. The objectives of the home visit are "to learn more about the personal and family background of the employee, to investigate the accuracy of the employee's personal information as stated in their personal data and to complement insufficient information presented by the employee."

A background investigator travels several times a week to different areas in nearby provinces to visit the family homes (but not the boarding houses) of workers who are on the verge of becoming permanent employees. In a typical day, the background investigator will visit seven to ten families. Upon locating the family's residence, the investigator asks to interview an imme-

---

5. In cases of maternity benefit claims, the BI staff contacts the National Statistics Office or local parish office to get a copy of the child's birth certificate. In the Philippines, the birth certificate shows the names of both the mother and the father and the date of their marriage. The BI staff then cross-checks the date of marriage with the date of hire and the personal information provided by the employee. Since Storage Ltd., as a rule, only hires operators who are single, they are able to check if the employee lied about their civil status at the time of hire.

diate family member, preferably the mother or father. Often, the house will be empty and the parents either in their fields or working nearby. During the time it takes to call the mother, father, or other interviewee, the background investigator begins filling out the Home Visit Report. A key area of the report is to quickly assess the socioeconomic level and condition of the family. The investigator notes the type of house (i.e. construction material used), number of bedrooms, sources of water and electricity, type of toilet, number and type of household appliances present, the total number of people living in the house, and finally, whether the house is owned or not.

When the mother, father, and other family members arrive, they are given few explanations, other than, "I'm from Storage Ltd. and we need to ask a few questions." In the fifteen or so interviews I sat in on, respondents appeared quite nervous and guarded, particularly when asked to provide detailed information on all family members, their activities, their occupations, and places of residence. After filling out the rest of the Home Visit Form with questions regarding educational background and previous employment (whether unionized or nonunionized), the background investigator also asked how the employee got the job at Storage Ltd. and where the employee now lives. In an almost off-handed manner, the investigator begins asking more probing questions, such as what the primary sources of income are for the family, whether the employee "helps out" financially, and how often the employee returns or sends money home. Then the investigator asks more directly, "How does your daughter like it at Storage Ltd.? Does she ever complain about it? Does she ever talk about any problems at work or disagreements with anyone at work?" These questions often make respondents intensely nervous, and few would give more than one-word answers, none negatively. Finally, the investigator usually asked who the local *barangay* captain was and how long he/she had been in office so the investigator could contact the *barangay* captain for a character reference. After the visit, the investigator makes further notes on the form regarding the visit. These include such comments as, "Parents uncooperative; very suspicious; seems to dislike home visit," or "The mother of [x] is very uncooperative and seemed to have an abnormal reaction regarding the home visit. Looks like [sic] *sumpungin* [cranky]."

The home visit, probably more that any other practice, exemplifies the depth and reach of Storage Ltd.'s surveillance systems. While designed to collect microlevel data on workers and their families, the home visit was equally effective as a way to directly intimidate employees and their families at the crucial period just prior to regularization. Workers and their families were made well aware that the company had tremendous power extending well beyond the factory gates and that they now had the intelligence to exercise it effectively.

Coerced Commitment

Managers at Storage Ltd. often talked about the need to increase workers' "commitment" to their jobs and company. But in contrast to much of the academic literature, the case of Storage Ltd. shows that worker commitment is influenced by a broad range of factors, positive and negative, internal and external to the workplace (Lincoln and Kalleberg 1996; Hodson 1997). As the previous chapter detailed, Storage Ltd.'s internal flexibility strategies do not depend on worker loyalty (since operator problem-solving is not required), and discretionary effort is achieved through technical control and disciplinary management. What remains important for a stable and reliable workforce is promoting worker attachment (Mueller and Boyer 1994).

Management at Storage Ltd. understands, probably better than anyone else, that their work system is in many ways very fragile: it depends on workers' willingness to consent to management's rather coercive strategies to maintain productivity. To "cloud" and protect this fragility and maintain control over both production and workers, Storage Ltd. opts not to increase pay, improve working conditions, or provide more promotion avenues but instead to play upon—and help solidify—inequalities in the wider labor market. Storage Ltd. uses a very selective and heavily discriminatory recruiting process to screen for vulnerability. Selection criteria based on age, sex, civil status, education, and experience were all designed to take advantage of segmented external labor markets. As Jill Rubery (1994, 53) writes:

> The existence of disadvantaged groups in the labor market is in part the consequence of broader social forces leading to discrimination within the labor market and elsewhere in the social system. However, disadvantage is also created through the policies of employers. Employer policies on recruitment, retention, and training all imply selection and selectivity. To the extent that employers' policies and preferences are similar and reinforcing, they necessarily contribute to the creation of disadvantaged labor market groups. . . . Individual employers are able to take advantage of this effect of the cumulative actions of employers within the labor market, and to use these disadvantaged groups to provide the stability and commitment required in organizations which lack the financial and institutional conditions to establish a primary (high wage and good promotion prospects) internal labor market system.

As will be discussed further in the next chapter on workers' responses and actions, many at Storage Ltd. may feel alienated by their actual working conditions, yet due to their backgrounds and the precarious labor market positions, are nevertheless more likely to be "attached" and committed to their jobs.

Integrated Production

Integrated Production has chosen to complement its flexible automation with an internal labor consent strategy that stresses positive incentives, communication, and individualized labor relations. For this reason, the firm's localization strategy is not nearly as deep or extensive as those at either Allied Power or Storage Ltd. However, their human resource management approach is still very much embedded in a selectively regulated local labor market and in many ways reliant on the anti-union tactics of the private zone developer and the interventions of state actors, both national and local.

In fact, Integrated Production was "born flexible," since even before starting operations, the company negotiated directly with the national government to gain legal exemption from Philippine labor laws (Article 83 on normal work hours and Article 87 on overtime work) in order to implement their compressed workweek policy. As a multinational firm planning to build a $200 million plant, the firm had enormous bargaining leverage with a host government hungry for such high-tech investment. Despite criticism voiced by the Department of Labor and Employment, two more powerful government agencies responsible for generating foreign investments, the Department of Trade and Industry and PEZA, eagerly supported the company's petition.

The firm also chose to locate in a private zone in the province of Cavite. As discussed above, Cavite has a notorious history of antiunion intimidation and violence. While much foreign investment has come into the government-run zone where Allied-Power is located, Cavite also has a number of private zones that have attracted some of the largest electronics corporations in the world, including Intel, Analog Devices, and Hitachi. These private zones, like the ones in Laguna province, provide state-of-the-art infrastructure, maintain strict security, and bar known labor organizers from entering the zone. For example, labor organizers that were successful in organizing in CEZ began targeting the zone in which Integrated Production is located. Yet after six months, the organizers had not even managed to get past the front gates, let alone to the firms within.

The private zone developer also plays a critical role in labor control and helping firms navigate local politics. Its explicit antiunion strategy is evident in its marketing materials, which tout the "absolutely strike-free environment . . . prevailing in Cavite province." This is supported by assurances that, "THE CALABARZON Police Assistance Group has a 16 man strong permanent outpost across from [the zone] to support its own security force within and around the vicinity of [the zone]."

These boasts are backed up by less visible actions. For example, from interviews with a local *barangay* captain it was discovered that he and other local official were on the payroll of the private developer as "civilian guards." The *barangay* captain bragged that he has personally known the developers since 1989, "before it was all developed. When the fields were just banana plants." According to another local informant, the title "civilian guard" is a euphemism for the developer's group of "armed goons." The informant noted that *barangay* captains, as the lowest-level elected officials, "make the best kind of goons" because they have the infrastructure, arms, and legitimate monopoly on the use of force. The *barangay* captain also has at his disposal on average ten *barangay tanod,* or community watches, who are then armed and paid through the municipality.

While at the time of the research, there had been no labor disputes within the zone where Integrated Production was located, there had been a strike at the new, seventy-two-hole, seventy-two-hectare world-class golf course across the road, which provided a glimpse of how local officials get involved in labor disputes. Laborers at the golf course, who were formerly tenant farmers who lost their land and were now hired to trim the fairways, went on strike to get back wages and be paid the legal minimum wage. During the strike, the *barangay* captains, armed *barangay* police, and regular municipal police got involved to "persuade" the workers to give up their strike. Research by a representative of the striking workers found that six *barangay* captains from the surrounding area were all "ghost employees" of the land developer and on the payroll of the golf course.

Integrated Production likely has neither knowledge of such labor disputes nor how they are handled. However, this does not mean that they do not benefit indirectly from the "industrial peace" that has been imposed in the province. The company's own strong antiunionism strategies that aim to substitute for a union are complemented by the more menacing and brutal antiunion tactics practiced in a province actively courting foreign investors.

Recruiting

In part because of its high quality standards and high capital investments, Integrated Production puts a strong emphasis on recruiting. Like many other electronics firms, it prefers women as production workers and hires no temporaries or contractuals. But rather than screening for labor market vulnerability like both Power Tech and Storage Ltd., Integrated Production has a lengthy and demanding screening process focused on education and "attitude."

The process begins with the basic hiring criteria. For operators, while the firm does look for young women ages eighteen to twenty-two, it also requires at least two or more years of college or technical education. This contrasts with Storage Ltd.'s strategy to cap the educational requirement for operators at under two years. While some departments prefer workers with no experience, other departments, particularly in production, want workers who have experience in electronics. There is also no stated policy regarding marital status, which is distinctly different than at both Allied-Power and Storage Ltd. where married workers are explicitly screened out.

Integrated Production also does not make an effort to recruit from rural areas. In fact, if anything, they have a preference for workers from more urbanized areas. During startup in 1996, national and local officials strongly recommended that the firm hire locals from the province and immediate vicinity. But the HR manager complained that the firm was constrained by requirements to hire only those from the province. She found that locals generally have a lower educational attainment level, tend to be the most unqualified, and often fail the exams. She added disparagingly, "Perhaps [locals] are better planting pineapple and coffee than working in a high tech plant." The manager then noted that the best operators are from the urban areas of the province or from Manila. For operators, the firm relies mainly on current employee referrals, sometimes from another operator, but more frequently from those higher up.

Clara, a test operator, explained the hiring process and the value of connections versus merit:

> My aunt gave my name, she's in accounting. Three hundred took the exam. In our batch, there were forty or fifty and only eight passed. Out of the whole group only twelve passed. We went directly to be interviewed the same day. I was the only high school graduate, all the rest were college graduates. There were four kinds of exams, mainly abstract reasoning. Then there are four interviews: first managers, then two test supers, then last is HR. The HR are strict. They ask me many questions, like am I against unions, and of course I answered in their favor, and I said, "yes sir."
>
> In our batch, the others that were college graduates who were approached by their backers, encouraging them, saying "Hi, you can make it." Me, no one approached me, so I thought I was dead, I'll never pass it, I'd better go home. But I made it. Some companies don't give a chance to the applicants. Only the ones that know insiders are picked, which is very unfair. Here, your backer gets your name on the list. After that, it's up to you whether you get accepted or not.

As evident, an unofficial but important screening criteria was to avoid those who might be sympathetic to unionism or those with a union background. One worker explained, "The company won't accept one that has come from a union. Like my older sister said. She's from Firm Z [a unionized electronics plant], and she said that anyone coming from Firm Z is not accepted at Integrated Production." Another operator confirmed this policy: "Someone from Firm Z applied [at Integrated Production] and they sent him home because they thought he was a union member. What HR did was to call Firm Z to see if he was on a list of union members."

The firm's particular hiring criteria have led to a slightly different workforce demographically than at other electronics firms. Due to a preference for experience, the workforce is slightly older: only 60 percent of workers are below twenty-four years of age. Also the older workforce and absence of marital status criteria has led to a workforce that is only 70 percent single. The demographic profile at Integrated Production is somewhat similar to that at Discrete Manufacturing in terms of age and marital status but quite different in terms of educational attainment. Integrated Production places a high premium on education and, according to the HR manager, 95 percent of the operators have attained at least two- or three-year technical degrees, while a large portion of workers are four-year college graduates. The education bias and the lack of provincial recruitment also contribute to the higher percentage of workers from urban areas.

Ambivalent Attachment

As discussed in the previous chapter, the firm's internal human resource approach to work organization has helped boost worker loyalty and effort. However, the firm's preference for educated women has had a somewhat negative effect on attachment and turnover. Their education and more middle-class, urban backgrounds mean these workers have more labor market options and are able to pursue alternative employment with similar wages in white-collar service work, such as teaching, nursing, banking, and retail trade. These educated workers also have higher job and career expectations, and often do not want to remain in production. In fact, many felt it is below their social and educational status to be "just an operator." Clara explained,

> They are strict and they want employees to be the best so you can contribute to the company, so they prefer college graduates. There are many who are college graduates, degree holders: there are BSN [B.S. in Nursing] and accounting, and there are teachers. It's a pity, if it were me . . . well, there's internal hiring anyway, that's what they are waiting for. If there's a vacancy up there, the company prefers them, hiring from inside.

However, the internal labor market cannot absorb all operators wishing to move out of production. In large part because their workers have more options and the threat of exit is more real, the firm tries to meet these expectations and reduce potential turnover by providing a friendly work atmosphere and a wide array of long-term, nontransferable benefits. Such generous benefits, what Becker calls "side-bets," build worker attachment because workers will be, in theory, less inclined to quit for fear of losing perks and supports on which they now depend (Becker 1960). While this strategy has been somewhat successful, the potential frustration of its well-educated workforce combined with the limited internal mobility may lead to higher rates of turnover as workers seek other jobs with a better "fit" to their educations.

Indeed, the turnover rate of more than 9 percent was higher than at either Storage Ltd. or Discrete Manufacturing. As we will see in the next chapter, many of these workers, who enjoy more labor market leverage and higher career aspirations, were not particularly attached to their production jobs, seeing them as either temporary, or as stepping-stones to other careers.

Discrete Manufacturing

Like Integrated Production, Discrete Manufacturing has focused its work regime and commitment strategies on internal firm processes. However, unlike Storage Ltd., Integrated Production, and most other companies in private zones, Discrete Manufacturing cannot use local and national government interventions and the zone developer to help guarantee a docile and union-free workforce. In large part because the firm is already unionized, the firm sticks primarily to negotiating positive incentives and does not have a fully developed localization strategy to boost worker commitment. However, these formidable constraints do not mean that management has not tried to use external strategies in the past to give itself more options in developing its work regime. Nor do current constraints prevent the firm from attempting new localization strategies to avoid previous "mistakes," or to limit the union's influence as the firm continues to expand.

The importance of external conditions and local embeddedness is clear from the firm's history and transfer to its present location. As mentioned in the previous chapter, Discrete Manufacturing has been assembling semiconductor chips in the Philippines since 1981. Their first chip plant, located in Metro Manila, was set up in a converted warehouse next to the parent company's other manufacturing plant. Crucially, workers in the firm's other plant had organized a union years before. When the new chip plant opened, workers from the older plant next door quickly helped organize a union, thus making Discrete Manufacturing one of the first and only unionized semiconductor plants in the Philippines.

By 1994 management wanted to expand, but they viewed Manila as an unattractive investment site because of the high cost of land, lack of industrial infrastructure, and relatively high rates of unionization. So Discrete Manufacturing, like Storage Ltd., opted to locate as the "anchor firm" in a new private zone in Laguna.

But while the facilities and much of the equipment were new, the bulk of production and almost half of its workers were transferred from the firm's unionized Manila factory. Not only was the entire union bargaining unit transferred to the new site, but it was also enlarged to include all new workers hired. The union now claims twenty-eight hundred members.

Recruiting

Despite the transfer of its older unionized workers, the firm needed to expand the workforce dramatically when the new plant opened and has tried to recruit a specific type of worker. First, the firm prefers new operators with at least two years of college and preferably, college graduates. They prefer women between the ages of nineteen and twenty-six years old with 20/20 vision. The firm also prefers female operators who are at least five feet two inches tall (five feet, six inches tall for male operators/technicians) and right-handed because of the machine sizes and design. They seek operators without previous work experience because, as the HR manager stated, "We want fresh people and we want to train them" in a firm-specific way. When possible, the firm also prefers to hire workers from more distant provinces or from rural areas.

To recruit such workers, the firm relies primarily on current employee referrals. But during the initial expansion, they also worked closely with the zone developer, local officials, and PESO to conduct outside recruiting. Through PESO, the firm was able to recruit in rural areas of Laguna Province and in the adjacent provinces of Batangas and Quezon. Victor, the head of central recruiting, noted that they preferred workers from the provinces because while in the local area, there were a number of zones, more opportunities, and a higher employment rate, in areas like Quezon Province, there are far fewer jobs. He also reiterated common conceptions about provincial workers as "shy," more willing to follow orders, harder working, and more serious. The HR manager even joked, "I hope the decline in agriculture continues" because it would mean further unemployment in rural areas and thus cheaper, more available workers for the firm. Finally, Victor commented that recruiting through PESO and local officials was mutually beneficial: the firm provides the local officials with a form of patronage, that is, jobs for local voters, but can rely on the local officials to recommend "problem-free" workers.

These strategic recruiting tactics are not unlike those used at Storage Ltd. Here, the same approach to rural recruiting to leverage worker vulnerability is present. However, this approach has had somewhat less of an impact at Discrete Manufacturing, given the existence of a collective bargaining agreement (CBA) with more built-in protections for workers and standardized, relatively high pay.

In terms of the gender balance of its workforce, Discrete Manufacturing is similar to other electronics firms: 75 percent of the factory workers are women. But Discrete Manufacturing workers are quite a bit older and more diverse. Among the 85 percent of the workforce that are factory workers, the average age is twenty-nine years old, with 77 percent twenty-six years old or older, and less than 1 percent were below twenty years of age. In part a reflection of their older age, only 65 percent were single. Workers interviewed did comment that the firm avoids hiring married women workers. Patricia, a twenty-two-year old single worker answered that no, they do not hire married women "because [they think] if you're married, you might already be pregnant and then you'll have lots of problems, like absences, and not being able to adjust to changing shifts." But while they try not to hire married women, because of their long tenure, many workers nevertheless get married, have children, and continue to work.

Attachment

Although high worker commitment at Discrete Manufacturing is primarily the result of the generous pay and benefits negotiated into the collective bargaining agreement, workers' assessment of their jobs and their attachment to them are also based on conditions outside the factory, and on their perceived positions in the labor market. For example, most workers first came to the job simply because they needed work and thus had few initial complaints. When asked why he chose Discrete Manufacturing, Edward, a twenty-four-year-old married male operator replied, "I just tried to get in. It was just more important to work than asking how much the pay was. Whatever the pay, it was fine for me." Once hired, workers often assessed their jobs in comparison with previous work. Aris, a twenty-six-year-old former seasonal sugar plantation worker noted, "The job's okay. It meets my basic needs. Of all my jobs, this one is the best, much nicer than the others." But particularly for the firm's older workers, many of whom have only modest educational credentials, security of tenure is a central motivator for staying. Dennis, a thirty-three-year-old married operator from the rural province of Oriental Mindoro who has been with the company for seven years, explained:

> I'm satisfied [with the job] because, somehow, I'm at least able to earn. Although I'm not satisfied with the wages. Still, I prefer to stay because it's really difficult to look for a job now, especially because I'm not a college graduate. If I didn't have this job, I'd be in the hills, farming.

Indeed, overall attachment to their jobs was very high. More than half of the twenty-four workers interviewed saw themselves still at Discrete Manufacturing in ten years and several workers saw it as their lifetime employment. Only three workers interviewed planned on leaving the firm within the next two years. Probably the greatest testament to the high level of worker attachment is the very low level of turnover. The average tenure of the twenty-four operators interviewed was 4.75 years. And according to HR Department records, the overall annual turnover rate in 1998 was less than 3 percent. Exactly why workers felt more attached to their jobs and how their high overall commitment relates to their nonwork lives will be discussed in the next chapter.

New Localization Initiatives and Antiunionism

The firm's rhetoric of worker empowerment and willingness to bargain with the union has not prevented the firm from trying to minimize the influence the union might have on further expansion. Indeed, during the time of field research, the company had just opened a new plant in another private EPZ in Laguna to produce more sophisticated integrated chips similar to those at Integrated Production. And in many ways, the management at the new plant modeled their work regime not on the parent firm's discrete semiconductor plant, but more along the lines of Integrated Production. When in full production, the plant is expected to employ nearly three thousand workers. However, this time, the parent company is trying to avoid the "mistake" it made when setting up the discrete chip plant in Laguna in 1994. Now, the firm is hiring an all-new, "fresh" workforce and not allowing any employee transfers from the current plant in order to avoid unionization. In comparing the two sites, the HR manager at the new plant stated simply, "It may be possible to work with unions, but we find not having a union better."

Management at the new IC plant has also tried a different localization strategy: that of manipulating and gendering a niche in the local labor market to reduce the bottleneck in—and bargaining power of—technicians. This more technologically advanced plant has reorganized work completely under its new automated production system. To support the new production process, the management is building a more skilled workforce made up mainly of new technical operators, similar to the ones at Integrated Production and Discrete Manufacturing. Consequently, the firm's hiring preference is for those with at least two years of college-level technical training. The HR manger, while excited about an entirely new and educated workforce, nevertheless lamented that it was difficult to find qualified women applicants for technician and engineering jobs. He claimed that women make better operators, but that changes in technology and increased automation are increasing the demand for technical competencies, and lowering the demand for manual dexterity. Still he wanted to hire women technicians in large part because, at least initially, they could be paid less than men.

To solve this "dilemma," the firm began working with several private training schools to train specifically women technicians. The schools, several of which are run by religious foundations, "train indigent people to become technically oriented." First, the schools source potential trainees from impoverished rural areas by testing recent high school graduates and offering full scholarships. The foundation pays for their training expenses while the firm pays a training stipend and provides five months of additional on-the-job training. The firm also helps write the training curriculum and donates machinery for the training school. In this way, graduates of the program become "perfect candidates" for work at the new plant, complete with firm-specific skills. As the HR manager stated, "It gives us an opportunity to see them at work, to gauge their attitudes and abilities better. We train their trainers so . . . we have much more input than a 'normal' selection process. We tell them what to look for. [The trainees] can be observed in how they relate with peers and their recognition of authority."

Because the new plant was still just becoming operational at the time of the study, it is still unclear what the effects the targeting recruiting will have on the gender balance at the new plant. The benefits and consequences of selectively training women technicians also remain ambiguous. On the one hand, providing employment and technical training to rural "indigent" women is providing both economic and social mobility to an otherwise marginalized population. On the other hand, the firm's manipulation of labor supply and selection of such vulnerable trainees was creating a workforce entirely dependent on the firm. Management's power over these new workers is only increased by the absence of the union at the new plant. And because much of the private technical training was so firm-specific, new graduates would not necessarily have wider labor market options.

Discussion and Comparison

Localization, Commitment, and Work Regimes

The case studies demonstrate that "global" multinational electronics firms are by no means entirely footloose, infinitely mobile, and divorced from the context of their production. As Collins (2003, 151) notes, "in every site where production touches down, it is instantiated differently. It must work through local institutions and establish the necessary web of social relations to get the job done." And in choosing how to "get the job done," questions quickly arise about the "spatial division of labor," or *which* area, with which mix of available labor, and under what terms might best facilitate production and capital accumulation (Massey 1995). The localization strategies of the four firms profiled in this chapter clearly confirm that "place" matters more than ever.

However, the cases also take the local/global debate a step further than simply addressing why firms choose *between* different locations or how they capitalize on uneven development (Smith 1990). As this chapter has shown, local labor markets are not just "given" and employers do not just exploit *existing* differences. Instead, the cases highlight how firms engage, manipulate, and try to reproduce local conditions and social relations to enhance production stability once investments are already sunk (Jonas 1996). This contrasts somewhat with Collins's (2003) arguments that "localization" strategies of global firms are necessarily positive and represent a commitment to a local community, while firms practicing "hyper-Taylorized" production tend to be more "deterritorialized" or detached from a locality. Rather, in the Philippines, localization strategies could be far from benign. In fact, because of the deregulated environment of the EPZs and the active role of the state, firms have the latitude to intervene *both* on the labor demand side of the labor market—in terms of the types of jobs they offer—*and* on the labor supply side—through highly selective recruiting of workers with particularly weak labor market leverage (Theodore 2003). Viewed from this perspective, strategic localization becomes a key element in the politics of both production and reproduction, or what I have termed the political apparatus of flexible accumulation. The four work regimes introduced in the previous chapter are more fully fleshed out here as they are combined with their different localization strategies, resulting in different types of worker commitment.

Thus localization strategies do not simply diverge, but vary systematically. Based on the four cases, we see that localization varies according to the character of labor control in internal work regimes, and each firm's use of positive or negative incentives to garner worker consent to that regime. In general, the greater the gap between the working conditions a firm provides and the level of worker commitment the firm hopes to generate, the more extensive the localization strategy needs to be in order to bridge that gap. And although both negative and positive incentives can promote commitment, a work regime using primarily negative incentives is at best going to elicit the weakest form of commitment: simple attachment. Nevertheless, firms that do not rely on complex commitment and workers' tacit knowledge for productivity and competitiveness might find a minimal level of commitment sufficient for stability.

Again, the Allied-Power group stands out from the other three firms for its less-flexible, lower-tech production and its willingness to use coercion both inside and outside the factory. But in part because Allied-Power favors direct control and eschews worker commitment, the despotic work regime is riddled with problems of reproduction. In fact, to sustain its internal despotism, Allied-Power turns to heavy-handed external strategies, exploiting slack and segmented labor markets to hire more dependent workers willing to put up with low wages, little security, and poor factory conditions. The firms also rely on the *selective* and often coercive regulation of labor by local actors, opting to invest in an EPZ and a province that promises—and brutally enforces—"industrial peace." But even with its external strategies and embeddedness in an elaborate local labor control regime, despotic production tends to lead to worker dissatisfaction and has proven neither stable nor effective in suppressing labor organizing, leading other, more advanced firms to avoid both public zones and the despotic route.

Storage Ltd. has a few characteristics in common with Allied-Power: it too uses an internal work regime that focuses on primarily neo-Taylorist production controls and chooses not to compensate workers with more positive incentives. For this reason, Storage Ltd. shares with Allied-Power one key attribute: a deep localization strategy and extensive political engagement with its local context.

But there are also important differences between the despotic and panoptic work regimes due to greater demand at higher-tech firms for flexibility, stability, and a minimum level of worker commitment. Internally, Storage Ltd. achieves dynamic flexibility through its technical workers and automated production, making the firm less dependent on having to offer positive incentives for worker loyalty or even effort. Yet for both production and workforce stability, the firm must still secure a third key element: attachment. Here, rather than providing positive incentives, the firm turns to external strategies, but ones different from those at Allied-Power. Externally, Allied-Power mainly *reacts* to labor conflict, using a coercive response to active labor organizing in tandem with local actors. Storage Ltd., on the other hand, requires additional stability and has thus chosen a more proactive, *preventative* approach—extending its gaze and disciplinary strategies far beyond the plant to thwart *potential* conflict and labor organizing. It also intervenes more directly and strategically into the labor market and into the lives of its workers, particularly through its targeted recruiting, background investigations, and home visits. Storage Ltd. thus combines both internal and external labor control strategies—greatly enhanced by the reorganization of the private zones and the selective intervention of both the national and local governments—to maintain productivity and stability, while minimizing labor cost, turnover, and discontent.

As will be discussed in the next chapter, selective recruiting for vulnerability has been quite "successful" for the firm. Given their gender, age, rural background, and inexperience, workers at Storage Ltd. often voluntarily consent to the factory conditions and pay levels because their jobs represent real improvement in their status and mobility. The firm has thus developed a stable work regime in which a group of particularly vulnerable workers has gained a measure of control over their own lives, yet not at the expense of management's control over the workplace, leading to the paradoxical condition of "coerced commitment."

The other two firms, Integrated Production and Discrete Manufacturing, also try to combine an engineer-led, high-tech flexibility with a stable workforce and uninterrupted production. Yet both the peripheral HR and the collectively negotiated work regimes use more positive incentives to either "purchase" or "negotiate" their workers' effort, loyalty, and attachment. This focus on positive incentives and a less dominating workplace means that neither firm must rely on a direct and extensive localization strategy. Yet the two work regimes and their localization strategies still differ, particularly in the level and character of worker commitment they actually generate. At Integrated Production, management closely models its work regime on its corporate parent, recruiting a highly educated workforce and using behavioral training and generous "side-bets" to individualize employee relations. However, the lack of genuine worker participation and blocked upward mobility, coupled with their preference for more educated, urban workers, has proven less successful in garnering worker loyalty and attachment. Integrated Production, while offering high-profile benefits such as a free computer, nevertheless also had the highest turnover rate of the three nondespotic firms. As will be shown in the next chapter, workers at Integrated Production feel that despite the positive perks, their jobs do not entirely live up to either the firm's lofty rhetoric or their own high expectations.

On the other hand, positive incentives have been central to the higher levels of worker loyalty and attachment at Discrete Manufacturing, measured in workers' positive assessments of their jobs and extremely low turnover rate. Here, despite still limited decision-making power in production, workers nevertheless retain the power to collectively negotiate positive incentives, security, and some mobility that reflect their own—and not simply management's—interests. But Discrete Manufacturing workers also consider external conditions in their attitudes and jobs assessments. These generally older and less educated workers recognize that it would difficult for them to get similar jobs on similar terms elsewhere, and are thus more committed to preserving both their jobs and their union.

Localization, Space, and the Transformation of EPZs

Despite differences in breadth and depth, all the firms, even the two firms with the shallowest localization strategies, are clearly locally embedded. For example, Integrated Production, although not directly involved, doubtlessly benefits from the antiunion practices of the local zone developer and local government in Cavite. Similarly, although Discrete Manufacturing has learned to live with its unionized workforce, management chose to expand operations in an entirely new zone, in large part because it wanted a truly "greenfield" investment that would allow it to preempt unionization, shape a different labor supply, and implement a commitment strategy not reliant on collective negotiation.

The variation in localization strategies thus highlights the importance of how firms reproduce the social relations in production, and do so with other actors and through space. It also shows that different internal work regimes require different modes of external regulation and a different character of engagement with a locality to promote stability (Peck 1996). Indeed, the vital importance of stability to foreign manufacturers and the Philippine state's own dependence on foreign investment mean that the state plays a critical role in providing the appropriate conditions for production and localization.

The clearest demonstration of the state's role is the reorganization of the export processing zones, which have been progressively decentralized but by no means de-regulated. In large part because the state's initial strategy of isolation and centralized despotism in the Bataan zone backfired, the zone authority changed its organizational tack in Cavite, adopting a neoliberal "rollback" strategy and shifting some regulatory power to the provincial and local level. Yet the local state's heavy-handed response to union organizing again led to instability and nervous investors. Thus in its third neo-liberal incarnation, the central zone authority has chosen a more preventive, "rollout" strategy, formally privatizing new zones, yet still coordinating zone placement, planning, and security (Brenner and Theodore 2002).

At the same time, the state has also been central in the political construction and selective *re*regulation of the local labor market. While some detailed studies of employer labor market strategies have argued that firms prefer to have their workers within close commuting distance, in the Philippine case this logic proved "disastrous" for labor control (Hanson and Pratt 1995). Instead, state officials, zone developers, and firms collude in a labor recruiting and control strategy that exploits social and spatial segmentations, which lead to group disparities in wages and work options. The state is also complicit in systematically undermining workers' bargaining power through its *inactivity* and *non*-enforcement of the Philippine Labor code, which guarantees workers' rights to organize and bargain collectively.

Finally, the state strategically interrupts the formation of working class communities through its active *dis*organization of worker housing. The use of boarding houses arranged near bus stops also allows firms to shift the costs of reproducing labor to other localities or workers' families, since workers can be more easily reabsorbed back into their rural, but proximate, provinces in times of downturns if they have not permanently immigrated.

Localization and Gender

The final element of strategic localization that all the case study firms share is the use of gendered recruiting strategies, exploiting women's weak labor market leverage in order to lower labor costs and increase attachment. But as with other aspects of localization, firms do not simply exploit existing inequalities, but also actively shape and reproduce them through their discourses and practices (Salzinger 2003). Chow and Lyter (2002, 49) point out that employers have an especially strong influence on shaping gender inequalities through their hiring and job assignments, making the workplace, "the structural embodiment of gender ideology." In the case studies, all firms and their hiring agents invoked the now familiar and near universal tropes about the "femininity" of electronics assembly work in ways that simultaneously tap and devalue women's work in order to maximize accumulation (Wright 2001). In the concrete process of assigning jobs and pay based on existing stereotypes, notions of "cheapness" and "docility" become gendered. As Salzinger (2003, 180) writes, "Women's disproportionate share of low-wage assembly work is due to the way in which their familial situation is understood, used and reconstructed by capitalist processes."

In the cases presented in this book, firms began by taking advantage of Philippine gender ideology embedded in the family structure, building on an existing association of women with reproduction and the domestic realm and exploiting their image as adjuncts to the "real" economy. In search of further worker "commitment," perceptive hiring managers at firms such as Storage Ltd., go so far as to tap the gender stratification within the Philippine family, specifically seeking oldest daughters because of their traditional role in supporting parents as well as younger siblings. Firms can also take notions of "appropriate" women's roles, say as caretakers, helpers, or supplementary income earners, then superimpose these gendered expectations on a range of new tasks and technologies (Chant and McIlwaine 1995; Parrenas 2001). In this way otherwise ungendered work can be labeled "unskilled" and deserving of lower pay simply because women happen to be assigned to it.

Despite this rather bleak scenario, it is important to recognize that although hegemonic gender ideologies clearly inform the work regimes at all the plants, the gendered meanings of high-tech work for workers themselves varied significantly. Indeed, workers responded differently to the diverse regimes and gendered practices—from quiet quiescence, to individual empowerment, to collective rage—depending on their own lived experiences and expectations. Thus struggles and reactions took many forms, often opening up a space for resistance and contestation. It is to these reactions and responses that I turn to in the next chapter.

5 Asymmetric Agency: The Workers Respond

"One Child"

I'm a child if I'm called
But awake to the truth
We are affected by violations
Conspiracy between government and bosses

Even though I'm just a child
I learned to join in the struggle for life
Wake up early, prepare my body
Because Mother is gone who helps me

What my parents always taught
Always work hard at school
So that when the day comes, I will not oppress
But will be the one who fights.

Hope to right the wrongs
That make it difficult for ordinary workers
Leaders wake up to the truth
That we the poor too have lots of needs.

CLETA PASCUAL, CEZ worker

SO FAR, I have focused in this book on the production and reproduction of labor control by a range of powerful actors with a stake in the high-tech factory: from supervisors and HR managers to governors and *barangay* captains. However, while employers and local officials certainly shape commitment, the voices of arguably the most important actors to this point have been muted: ultimately it is the electronics workers themselves that consent to or resist particular labor regimes, constructing their own identities and forging their own meanings in the process. Indeed, while firms develop hegemonic forms of control to generate worker consent, these actions—and the often wide gap between management rhetoric and workers' lived realities—are also key sources of contradiction, conflict, and contestation. In this chapter, I bring workers' voices and experiences to the fore, drawing on interviews with workers in their workplaces, homes, and boarding houses to trace their gendered experiences of high-tech work, the meanings they attach to it, and their active responses and struggles. As will become quite clear, even under repressive conditions, workers are by no means passive. In fact, as events at Allied-Power and the Cavite EPZ will demonstrate, workers can find innovative ways to act, taking advantage of employers' need for labor's consent to develop their own localization strategies to boost their collective bargaining power. Workers' reactions to their jobs and work are also diverse, stemming in large part from their respective employer's labor control regimes. But influencing their divergent responses were also factors far beyond the shop floor, such as workers' own backgrounds, their labor market expectations, perceived social status, control over income, and familial responsibilities. Thus although the pressures of the high-tech shop floor are often unrelenting, factory work can still deliver real benefits and increased autonomy outside of production, which workers prize quite highly. For women workers in particular, the economic, social, and cultural capital earned through their clean, modern, high-tech work may help them challenge other forms of everyday subordination in their families and communities.

Nevertheless, workers necessarily help constitute the labor regimes they consent to or resist. In spite of the benefits of high-tech work to workers' personal lives, without collective organization, such individualized or "asymmetric agency" does not challenge management authority in production, thus demonstrating how workers' actions and discourses can simultaneously challenge *and* reproduce their own subordination and capital's flexible accumulation strategies.

The Squeaky Wheel's Dilemma: Unionization at Allied and the Rise of a Provincial Labor Alliance

> In a labor surplus economy like the Philippines, the squeaky wheel doesn't get greased, it gets replaced.
> RICHARD SZAL, ILO director for the Philippines

As the previous two chapters have demonstrated, workers at Allied Ltd. and Power Tech face pretty grim realities. Both firms chose primarily despotic work regimes to bolster their "low-road" flexibility strategy, based on minimizing all costs and adjusting—or simply downsizing—their workforces to match short-term demand (Deyo 1997). Internally, they tightly control and sweat workers. Externally, they rely on a network of local actors to suppress collective organizing, while at the same time they recruit workers with few labor market options. Finally, when even these tactics prove insufficient, the firms can simply shut down production or threaten to leave.

Facing concerted labor control, the mobility of foreign capital and slack labor markets it would seem that Allied-Power and other EPZ workers seeking redress for their injuries do not stand a chance. This is also the view prevalent in much of the literature on EPZs, which has focused on the multiple vulnerabilities of young women workers (Elson and Pearson 1981; Stichter and Parpart 1990), the state's active role in labor suppression (Deyo 1989; Southall 1988), and the hyper-mobility of capital (Brecher and Costello 1998; Burawoy 1985). These themes of exploitation and powerlessness have only intensified in the writing on globalization, which argue that increased international capital mobility automatically trumps actions by local labor (Boswell and Stevis 1997; Greider 1997).

Yet as noted earlier, despotic work regimes can face crises of reproduction and can clearly generate their own instabilities. Below, I trace the unlikely but (partially) successful unionization by the workers at Allied Ltd. and their innovative organizing tactics and alliances. Here, workers and their local allies do not have the means to "go global," as much of the antiglobalization literature suggests, trying to match the mobility of firms with their own "counter-hegemonic globalization" (Evans 2000; Moody 1997). Rather, they chose a primarily *local* strategy, developing a novel or hybrid form of what Herod has called "labor's spatial fix" (Herod 2001). Recognizing labor's ability to create its own localization strategies is to give workers a much more active role and to highlight how workers themselves use and create space in their own reproduction.

Workers' Responses at Allied Ltd.

To understand workers' actions and assessments of their jobs, it is important to first put them in the context of Allied Ltd.'s labor control and recruiting strategies. Clearly, Allied's despotic work regime does not rely on eliciting dynamic worker commitment, but instead seeks simply compliance. To achieve this, the firm relies heavily on their external strategy of hiring young or even underaged women. As noted in chapter 3, 80 percent of Allied's production workers are women, nearly all of whom do not have an education beyond high school. Although hiring underage workers is common, these young women have severely restricted labor market options, since they cannot legally work in the formal sector in Manila or in the larger electronics firms in Cavite or the neighboring province of Laguna. For most, work at Allied was their first jobs and the first time they were earning an independent wage. Although they were often being paid below the minimum wage and less than others even in the CEZ, the wages were above those they could earn in the urban informal sector or as agricultural laborers. For example, Angel recalled how she felt when she had just begun working in 1992:

I was happy because I was earning. I could buy clothes, makeup, etc. When I was with my parents, I couldn't depend on them [for such things]. I mean, they're just farmers. So I was happy. It was the first time I earned 87 pesos a day [approximately three dollars], so of course, I was happy. In fifteen days, 1,050 pesos. I remember that 1,050 pesos. I was very happy.

At the other end of the short age spectrum, many of those at Allied who had not quit after the extreme slowdowns were "older" workers, that is, over twenty-four. Even when they were called in only one or two days a week, or even sent home for two to three weeks at a time, most stayed because they had fewer labor market options, making them more dependent on their jobs at Allied. When asked what she would do for work if Allied finally shut down, Angel, who had recently turned twenty-four, replied: "For me, I would like Manila, or Laguna. I'd like to do the same thing, electronics. If I can still qualify, age-wise. There's an age limit, until you're twenty-four or twenty-five, then you can no longer work there. I don't know if they'd take me anymore."

Other workers felt "attached" to their jobs in part because they supported not only themselves, but also other members of their family. For example, many women workers who have younger siblings are asked by their parents to help pay school fees. But paying the college tuition for multiple siblings is generally beyond the budgets for most Allied workers. Particularly as work (and thus paychecks) have been so erratic, some are obliged to borrow at usurious rates to fulfill their family obligations and keep siblings in school. When asked if she contributed to her family's expenses, one worker stated:

I can still give some . . . to my siblings. I send some of them to school. In 1992, only two were in high school and I could still support them. I needed to give two thousand pesos every month. But in 1997 and 1998, they started college and to support them, I had to get cash from the office, from accounting. But it's "Five-Six."[1] Now, I can hardly give. My sister is down to one semester because my work has been affected.

In fact, it is a great irony that the firm's strategy of exploiting workers' weak labor market position to create dependence has worked too well: those left at Allied have few other options, so have chosen not to quit in hopes of winning severance pay based on their length of service through the union.

---

1. As noted in the previous chapter, "5–6" interest rates refer to the interest charged by local moneylenders: for every five pesos borrowed, six pesos must be paid back. The "5–6" rate is equivalent to 182 percent interest a year.

Meanwhile, the Allied management, through its slowdowns and job rotation strategy, is simply trying to find a way to get rid of workers without have to pay retrenchment fees. Some workers, like Lisa, have been with the company since they began operations and despite the working conditions, have demonstrated not only simple dependence or attachment, but real discretionary effort and loyalty. However, Allied's tactics have only left such workers feeling angry and betrayed:

> I was a line-leader before, in charge. That's where I put all my time, I came in at six A.M.—we actually started at seven thirty or eight—to get all the paper together that needed to be signed by the managers. And before dismissal, there was OT. The one in charge of me in the department, she wanted me, before going home, to complete the paper work, so she could just sign them in the morning. She wanted me there, after the "ding-dong" to do the reports. And there was no OT [overtime pay]. Sometimes I was there until nine P.M. and many times, I was all alone. There were four office staff with me, all supers [supervisors]; I was the only operator. I had to submit all the weekly and monthly reports. Me, an operator. There were no Sundays and happenings for me. I brought home all the paperwork. I did all the work at home. I gave my all to them, that's what hurts.

After retrenchments reduced the workforce from 750 to 250 and the firm started forced vacations and rotations in 1997, several workers sought out labor groups that were just beginning to resurface in Cavite for advice about union organizing and legal recourse. Rose, one of the new union officers, explained:

> We heard they would throw us out of production, so we talked, two of us. We made a plan to bring the workers to [union A]. . . . We decided go with [union A], because we knew someone, Paul. He lived next door to Eve, and he explained things to us. So I learned what our rights were. At first, we were only two. Then Paul said, "Bring more people." So we got eighteen for our second visit. Then on our third, we had twenty-four, and [on the] fourth, thirty-seven. On the fifth, we had our GE [general election]. That's how we triumphed. . . .

But then she went on to explain the management's reaction:

> That's how we made it so fast, but it got dissolved just as fast, because our officers were paid. The company offered immediately. . . . I went to the one I suspected was approached by the company, to see if it was really him. And it was! Also the vice president. . . . They made an offer to me, but they couldn't buy me, they couldn't force me, they know what I am.

In fact, the company strategy was to approach all the *male* union officers and offer to pay them individual "voluntary separation pay," really bribes, if they would quit the union and give up their demands that the company provide separation for *all* workers. The male officers accepted the payoff and immediately quit, prompting management to petition for union decertification with the Department of Labor and Employment (DOLE). In part because of pressure from the zone authority (PEZA), the DOLE did threaten to decertify the union on a technicality, forcing the remaining workers to hold another election. The election led to a new union with all women leaders, which has so far resisted individual payoffs by the company.

Labor's Scalar Fix

The nascent union at Allied would likely not have survived the concerted union-busting attempt if not for the material and moral support from other workers and organizers in the province-wide Cavite Labor Alliance–CLA (pseudonym) labor alliance to which they belonged. As one labor organizer from the alliance put it, "Our main goal is to organize workers, since everyone else they face is organized: PEZA, the local government, the personnel managers, all of them. It's only the workers who aren't organized."

The zone-based labor alliance began quite humbly as an outgrowth of a church-based sociopastoral program. However, in only a seven-year period, it grew to become an important force in the zone with twenty-two unions and affiliated groups with more than two thousand union members. It would also serve as an important model for organizing successful *local* struggles against multinational firms and their state allies. The CLA sprung from the initial actions of a local Catholic priest, who had worked with labor groups and organized basic ecclesiastic communities at his previous parish in a nearby town. When he was assigned to serve the Most Holy Rosary Parish in Rosario, which happened to include the worker communities surrounding the zone, he brought with him an interest in worker issues and a church-based, community-organizing model to address them. But at the time of his transfer in 1995, the climate of fear was still quite strong, with former governor Remulla still retaining the loyalty of local leaders who provided the microlevel muscle to intimidate union organizers. The priest set up the Rosario Workers' Center (RWC), a small space adjacent to the church and began organizing nonpolitical events such as prayer meetings, a Saturday night workers' mass, and a zonewide choir. He also organized social events like beach outings, discos, and birthday parties to provide a needed break from the monotony of factory work. These events served as a nonthreatening way to bring workers from different factories in the zone together, provide a venue to air workplace grievances, and introduce the organization and workers' issues. With more than forty thousand workers in the zone, most of whom live in the crowded boarding houses crammed in the immediate vicinity, the church's activities were also aimed at raising workers' class consciousness and building a strong local, community identity among the atomized, mainly migrant-worker population.

The Rosario Workers' Center's small staff did not have direct background in labor organizing, but were mostly women with experience in church-affiliated urban ministries, progressive peasants organizations, and the wider women's movement. What these organizers had in common was participation in community-based groups that played key roles in the anti-Marcos struggles in the 1980s. At the center, they established livelihood programs, a workers' loan fund, and provided legal assistance to help workers file grievances with firms, the zone administration, and the Department of Labor and Employment (DOLE).

Importantly, the workers' center specifically addressed women workers, who make up more than 73 percent of the zone workforce. The center often framed its appeals in terms of a gendered worker identity, publicizing cases of sexual harassment at work, the firing of pregnant workers, discrimination against married and older women, exposure to hazardous chemicals, and the prevalence of urinary tract infections among workers due to forced overtime and prohibition of toilet breaks during shifts (WAC 1996; WAC 1999; WAC 2003b). Cavite Labor Alliance that has grown out of this initial organizing has tried to maintain this gender balance among its organizers and leadership: the president of the labor alliance is a woman, as are over half of the fifteen members on the leadership council.

In 1997, the direct repression of union organizing in the province had subsided under the new governor, and several labor federations were actively organizing in the zone. The worker center took the opportunity to launch a new entity, the Christian Labor Alliance, a nongovernmental labor organization (NGO), created as a workers' organization with a direct church connection, which nevertheless would appear less threatening—to both workers and the government—than a traditional trade union. By this time, the workers' center and the NGO had established enough of a presence in the zone that workers were beginning to seek them out to address workplace grievances. It is at this point that two Allied workers approached Paul, one of their neighbors and a labor center community organizer, to help them begin forming a union.

By 1999, the Rosario Workers' Center and Christian Labor Allicance had overcome most of the early difficulties of starting up, becoming recognized and legitimate advocates of workers' rights in the zone, with close working relationships with other community groups. In particular, they had developed close relations with the Federation of Nongovernmental Organizations (FENGOR), a multisectoral collection of forty other NGOs in and around the town of Rosario that worked primarily among the urban poor, women, fisherfolk, and landless agricultural workers. These close relations helped make labor problems not simply a workplace or even a zone-specific issue, but as a local community issue, which put more pressure on the local government to remain neutral and not intervene during labor disputes.

However, as vital as local community connections are to keeping firms and their local allies in check "from below," organizers also recognized early on the importance of international networks that could bring added pressure "from above." Thus, as early as 1996, they had forged ties with a number of progressive church, labor, antisweatshop, and trade union organizations around the world. These organizations provided some financial support, allowed them to conduct and publish—in Filipino—the results of a survey of zone conditions, and also helped them participate in international workshops involving women EPZ organizers from Asia, Latin America, and the Caribbean. The international networks also helped support local organizing: for example, by bringing international antisweatshop letter writing campaigns to bear on firms, local officials, the DOLE, and the PEZA.

By 1999, the labor center and NGO successfully helped establish nine independent unions at the shop floor level, including the union at Allied Ltd. However, as a church group and labor NGO, they could not formally be involved in representing workers in labor negotiations with firms. Thus the organizations transformed themselves once again, this time into the province-wide Cavite Labor Alliance, with the ultimate goal of becoming a labor federation.[2] For the first time, organizers campaigned openly as a militant, rather than "Christian" labor alliance.

By 2000, the now openly militant Cavite Labor Alliance had developed a dense network linking local workers and labor organizations and trade unions worldwide. Their efforts were significantly strengthened by the heightened awareness internationally about conditions in the CEZ brought about through Naomi Klein's (2000) antisweatshop book, *No Logo*. Building on this momentum, the provincial labor alliance launched a concerted international campaign in 2001 to support local actions focusing on labor rights violations and antiunion activities in nine firms in the CEZ, including Allied Ltd and Ultra Electronics. The campaign also highlighted the complicity of the CEZ administration and the local government in labor and union suppression. The international campaign targeted action at several levels, including multinational firms with ties to the CEZ, the Philippine president, the national zone authority, the DOLE, and local officials.

Here again, the workers at Allied benefited from the alliance's actions and networks. As noted above, the Allied management attempted to dismantle the workers' initial union in early 1998, largely by offering bribes to male union officials to drop their demands for the firm to pay all workers their legally entitled separation pay. However, these attempts at union-busting proved unsuccessful as women workers stepped into the leadership roles. But production remained erratic and in 2001, the firm again nearly stopped pro-

2. In order to become an official labor federation, the group must have at least ten affiliated unions with established collective bargaining agreements (CBAs).

duction, with workers being called into work as little as two or three days per month. It was hoped that workers would not be able to survive under such conditions, and they would simply abandon their demands and quit. Management also tried to bribe the female union leadership with individual offers of voluntary separation pay. However, the provincial labor alliance responded by making the Allied case one of their core campaigns and sending appeals around the world calling for allies to petition PEZA to revoke Allied's business license for nonoperation and order the firm to pay all back wages and separation pay. In part because of international and local pressure, PEZA ordered the company to stop offering the voluntary separation pay scheme, officially recognize and begin bargaining with the workers' union, and resume normal operation of the company. Faced with the mandate, the firm reluctantly recognized the union and subsequently resumed operating four days per week, but did not guarantee that such steady work would continue (WAC 2001b).

By 2002, the workers' center had at least six staff members in Rosario and three each in two satellite offices near several private EPZs in the province, which provided education, trade unionism training, and legal assistance to workers, while also maintaining network ties to international solidarity groups. The actual labor organizing arm of the alliance had fifteen organizers, but only four were full-time. In all, the Cavite Labor Alliance claimed fourteen official unions, eight affiliated chapters (with unions in different stages of formation), and twenty-four additional factories at the contact and initial organizing stage. They also claimed that slightly less than twenty thousand workers have been directly influenced by and benefited from their myriad services over the previous seven years, with more than 60 percent of these workers at the Cavite EPZ.

The case of the Allied workers' unionizing drive and their affiliation with the Cavite Labor Alliance brings out three important factors that help account for the rise in successful local union organizing, despite the reach and complexity of the despotic work regime and the complicity of the local government. First, because systems of labor control rely so heavily on extrafactory regulation, political changes at the subnational and local level can help create the political opportunity for organizing. Second, the most successful labor organizing strategies may be local and community-based, paying particular attention to community gender issues, rather than factory or industry. As labor control strategies have extended outside the factory, labor groups have countered by expanding their own strategies that appeal specifically to women workers and drawing on the progressive traditions and organizational resources of earlier, community- and church-based social movements. Third—and probably most innovative—the case of Allied workers and the Cavite Labor Alliance demonstrates the importance of taking multiple organizational forms and strategically extending action repertoires to multiple geographic scales, what might be best described as labor's "scalar fix." Navigating between the pressures of globalization and their vulnerability to what Tarrow (1998) calls "the tyranny of de-centralization," the Cavite Labor Alliance was able to shift its institutional form and appeal: from church group, to labor NGO, to trade union alliance—and sometimes back again—in order to leverage the advantages of these different institutional identities.

The organization also forged network ties at multiple levels. Making use of their deep connections with local community groups within and outside the zone, but also drawing support from national-level unions and tapping the influence of international labor activists, the alliance brought diverse and multiple pressures on firms and government agencies and gathered support for the alliance's core efforts to organize local workers.

Work versus Earning a Living at Storage Ltd.

Like Allied-Power, Storage Ltd. relies on an extensive localization approach but follows a more nuanced strategy, creating the conditions in which workers *voluntarily* commit—rather than simply submit—to the rigors of high-tech factory life. Storage Ltd.'s tactic of recruiting vulnerable and inexperienced women workers, but providing relatively high take-home pay due to daily forced overtime, has helped the firm minimize costs while maintaining stability and forestalling any collective worker organizing. But if the panoptic work regime has clearly proved "successful" for management, what are the benefits for workers themselves? How do they view the firm's control system and their own complicity in it?

Below, I present the panoptic work regime from the workers' perspective, showing that despite the constraints in production, these women have managed to achieve some measure of autonomy in their nonwork lives. In this way, Storage Ltd. workers seems to echo the experiences of women workers in other parts of the globe, who have also benefited from low-wage work: not necessarily in production, but from their general labor force participation and wage earning (Wolf 1992; Lee 1998; Tzannatos 1999; Freeman 2000; Chu 2002).

However, in reviewing the lived experience of Storage Ltd. employees, it must also be remembered that they represent a select—and selected—group of workers. Indeed management carefully recruited them precisely because it presumed that their weak labor market options would make them both more attached to the jobs and more grateful to simply be employed. As noted in the previous chapter, 88 percent of the firm's workers are women, 97 percent are single, 98 percent have less than a college degree (49 percent have a high school diploma while 42 percent have some college or vocational training), most are from rural areas, and their average age is twenty-one. Thus, while there is no doubt that the women workers benefited from their jobs, when looked at broadly, their weak bargaining power vis-à-vis management is a conscious product of the firm's supply-side labor market interventions that exploit persistent social, economic, and spatial inequalities in the Philippines. As Elson (1999, 618) notes: "The increased participation of women in the paid labor force can represent 'distress sales' rather than free choices to take up new opportunities."

In order to get a better understanding of workers, their perspectives on Storage Ltd., and their responses to the work system, eleven Storage Ltd. workers, ten women and one a man, ranging in age from twenty to twenty-three were interviewed. The man, a warehouse operator, is married, as is one woman, a trainer. Eight of the nine single women are production operators, and one is a newly hired technician. Nine of the eleven interviewed live in boarding houses or shared housing away from "home." Respondents are originally from provinces as far off as Panay, Negros, Marinduque, Mindoro, Manila, Quezon as well as two from Laguna. Five of the respondents' parents were farmers/fisherfolk, while other parents' occupations ranged from unemployed, to guard at a local school, to government worker, to small businessperson, to sales representative. Nearly all had some college or were vocational school graduates and had surpassed their parents in educational attainment. For eight of the eleven workers, Storage Ltd. was their first employer.

Similar to the women in other studies of EPZ employment, many sought jobs at Storage Ltd. in order to escape rural poverty, find jobs in the better-paying formal economy, and "help out" their families (Chant and McIlwaine 1995; Lee 1998; Chu 2002). When asked why they took jobs at Storage Ltd., most cited reasons such as "pressure from home and financial crisis" and "pressure to start working [earning]."

As Hodson (1997) theorizes—and Storage Ltd. management clearly understood—workers' attitudes about work, their job satisfaction, and their behaviors on the job are clearly influenced by external factors. Because workers were often selected for their *in*experience and few had previous work histories, they often compared their situation at Storage Ltd. with the work situation of friends and family. Nina, a twenty-year-old production operator, was recruited from the distant province of Panay for work at Storage Ltd., her first job. She now lives in a boarding house with workers from Storage Ltd. and other factories:

My job is pretty good. It's not real heavy work. The important thing is I earn money. The only thing that's really hard is the OT. . . . But I'm satisfied at Storage Ltd. It's not really enough [pay], but it helps anyway. Compared with Robina [a nearby food processing plant where her boarding room mates work], its nicer at Storage Ltd. because they're all contractuals there [Robina]. There are more benefits at Storage Ltd. too. It's better because there's OT. . . . The pay is a lot if there's OT. But it would be nice if we could have a day off when we asked.

Workers also evaluated their situation by comparing it with the bleak, alternative labor market options they faced. When asked what they would do if they lost their jobs or didn't get them in the first place, many said they would simply stay in the province, farm, or look for whatever work they could find. A male worker noted: "In my situation, . . . my condition now isn't too bad. A lot of my classmates from college who have already graduated, almost all of them have no work. They're just unemployed. If I wasn't working at Storage Ltd., I'd probably just be a jeep driver."

Geraldine, a twenty-two-year-old operator and first-time worker, also felt her job was okay and had no immediate plans to leave: "I'm happy and the work is not too heavy. . . . If I didn't have this job, I'd probably just stay at home since my parents can't pay for my studies, although I'd like to keep studying."

Others, who had been in the formal labor force before, viewed work at Storage Ltd.—and in the electronics sector more generally—as both more stable, cleaner, and more prestigious. Kiko, a twenty-two-year-old with previous work experience, elaborated:

Before, I worked [at a biscuit factory] as a blender operator for two months. Then I worked at [a car parts factory] as a press machine operator in the casting department for five months. Before I quit, I was already accepted by Storage Ltd. The work now at Storage Ltd. is nicer [*mas maganda*] because at the first two jobs, it was physical work. Now it's more mental, though there is still some manual work too. At least, when I'm making reports, it's less tiring [*kahit paano ay nakakarelax*]. My old work at the car company didn't seem too secure. But the computer industry is more stable because this is now the computer age and Storage Ltd. even has worldwide operations.

Interviewer: How do you like your job?

I like it pretty well. Sometimes, its kind of hard—for example, you're not quite finished with what you're working on when the next thing to be done comes up right away. It's especially hard during peak season, though I'm glad [that there's so much work] because it means the company is doing well [*malakasang kumpanya*].

Others also echoed Kiko's view. Graciela, one of the few female technicians, said:

Yes, I like this job. It's got high status. I haven't been looking for any other work. . . . There's eight thousand workers already in the company. They're even hiring more. And there's a new building going up in Batangas. . . . The work is okay. In terms of training . . . if you're at Storage Ltd., you're really trained well. They look at everyone the same, even if you're female, and you really get trained.

Interestingly, Kiko and Graciela, who gave the most positive assessments of the company, were both nonproduction workers and commented positively about conditions for them within the factory. Both noted how hard work in production was and how glad they were not to be on the assembly line. They also saw themselves as having access to an internal labor market and upward mobility in the firm. When asked how far they thought they could rise in the company, Graciela answered the highest level: supervisor. With her entry position as a technician with a technical degree, she did in fact have this avenue open. Kiko, who hoped to return to his studies to finish his degree and who had an inside "backer" at Storage Ltd., also felt he could rise to a management position. Their positive assessment of Storage Ltd., then, is almost certainly colored by their real prospects for upward mobility.

The other nine workers interviewed were all operators and most stated they could rise only to sub- or group leader and several stated that they could only remain operators. These respondents had a less rosy view of the company and their prospects for moving up, but nevertheless tended to consider themselves lucky to be working at Storage Ltd.

Storage Ltd. management's strategy to hire "breadwinners"—that is, oldest children responsible for supporting the family—was only partially successful. As one worker noted,

Before I was married, I used to give my entire salary to my mother [*iniintrega*]. I only left myself an allowance. Now that I'm married, my parents don't force me [to give them money], but I do anyway because they lost their jobs and so I've become the breadwinner for the family. When I got married, it was okay because my sister was already working. When I was still studying, I had a fight with my parents because they wanted to take out a loan so they could still pay for my schooling. I insisted that I go to work because [my parents] were so poor. My father was upset and against me working.

However, while some workers felt "obliged" to support their parents and siblings, many did not actually sent money or return to the provinces, in part because they could not afford to, but also in order to keep control over their income and assert their new-found autonomy. A number of workers stated they only remit money home infrequently, or simply "when I feel like it." Jenny, a twenty-year-old first-time worker who was also directly hired from the province said, "I give them [her parents] money when they ask, but not regularly. I send P3,000 twice a year; but it depends on how big my salary is. I only see my parents occasionally: once a year."

Others said they simply did not have enough left over to send money home. Several workers noted that their base pay was far too little to live on, and that only their overtime pay made it enough to get by. In fact, many workers were forced to go to moneylenders to make ends meet. It was quite common for workers to "pawn" their ATM cards. Under this popular scheme, workers give a loan shark their ATM card in exchange for the loan amount, say P10,000. The lender then uses the ATM card to draw the worker's salary directly from the bank, since all salaries are paid by direct deposit. The lender takes out P1000, then "pays" the worker the balance of her monthly salary. This goes on until the loan and interest—at a rate of 182 percent per year—are paid in full.

When discussing their actual jobs and the amount of time their jobs consumed, workers also expressed their dissatisfaction. Nearly all felt that the overtime was overwhelming and complained they had little time for themselves or their families. Kiko, when asked what he did in his spare time or if he joined in any social activities in his community said: "I don't belong to any groups around here. I'd like to, but I don't have time. I don't even have time for myself! When I was younger, I was chairman of the Youth Group [*Sangguniang Kabataan*]. But now that I'm a breadwinner, I don't have time." Similarly, when Nina was asked if she was active socially in the community she commented, "I don't have time to hang out. If I have any time, I just sleep."

In discussing conditions at the factory, workers complained of headaches, eyestrain, backaches, and most commonly, "overfatigue" from the pace of work, coupled with insufficient rest. Remember, one of the most common offenses recorded by the HR department was getting caught sleeping on the job. Three workers also explained that they have had urinary tract infections (UTIs) and that these were very common among Clean Room workers, due to the short breaks. As Emmy noted, "Our break is only ten minutes, so there's no time if I'm in my bunny-suit; it's too difficult." Another worker shared a similar story, "The most common illness? migraines and UTIs. I handle a microscope [in the Clean Room], so I just hold it in because it's too much trouble to take on and off my uniform." Workers are given only two ten-minute breaks and a one-hour lunch break over the course of eight and sometimes twelve hours.

Two workers, who at the time of the interview were working seven days a week, felt that the situation was quite bad and that a labor union could help. Irma commented, "I would like a union because if there was a union, the workers would be protected. For example, it could help fight OT. There's just too much OT."

Belinda had a similar response; "We don't have time for ourselves because of OT. There's not even time to do laundry. So yes, I'm for a union. But the only thing is, other workers are afraid because they might lose their jobs."

This fear was borne out in the other interviews. The other nine workers did not support unions, thought they only made trouble [*magulo*], or did not feel one was necessary at Storage Ltd.

Despite some strong feelings about their lack of time and rest, workers also noted that their jobs at Storage Ltd. did give them clear benefits. Working and earning wages allowed many of the workers a measure of independence from their families and freedom from rural and provincial life the first time. Irma, quoted above, said: "My father wanted me to become a teacher, but I didn't want to teach. I wanted to study engineering, but I couldn't afford it. . . . Our parents didn't push us to work. We wanted to work. If I'm going to study, I want to be the one who pays for it."

Others also experienced a rise in status within their families. Jenny, the youngest of eight siblings, had a generally good assessment of work at Storage Ltd. and related it her status in the family: "This is my first job and I'm satisfied with the pay and benefits. I have no plans to work abroad; I'd rather just work here. Working for twelve hours is not so bad. The work's pretty light . . . my parents treat me better now that I have already have my own job."

Interestingly, there are some instances in which company policy benefits both the company and the women workers, but for different reasons. For example, as the firm rapidly built up its women-dominated workforce, it experienced what one staff member called "an alarming trend in the number of pregnancies." For the firm, this trend meant paying for up to six weeks of legally mandated maternity leave. As the workforce grew to more than two thousand in 1996, the number of workers who applied for maternity benefits jumped from nearly zero to 78. In 1997, as the workforce more than doubled to forty-five hundred, the number of pregnancies jumped again, to 180. At this point, the firm launched its Reproductive Health Services program within the company, giving workers access to family planning services and contraception. In 1998, as the program went into effect, only 203 pregnancies were reported despite another doubling of the overall workforce.[3] In the heavily Catholic Philippines, where contraception can be difficult to obtain and where divorce remains illegal, this access is often quite welcome by women workers. In fact, during an Employee Relations meeting, operators asked if they could invite their spouses to the company's Responsible Parenthood Program, to which the management readily agreed.

Finally, workers were asked how long they planned to stay at Storage Ltd. and what their future plans were. Some, like Anna, a twenty-year-old operator, were somewhat resigned: "I think I'll just stay there. I'm a regular now

---

3. But the low level of *reported* pregnancies might also be due to workers fear of losing their jobs.

and it's tough to look for work. . . . If I weren't working here, I'd be back in the province [Mindoro] helping with farm work."

Irma, an operator in the slider section, had a similar comment: "I got lucky to land there and its okay. I like my co-workers. I don't want to stay, but I promised myself I'd stay [at Storage Ltd.] five years. After five years, I'll go to the province. I'll set up a business. Having a business, it's easier to earn more money. I want my own hours."

Seven of the eleven interviewed saw themselves still at Storage Ltd. in two years while three saw themselves with the company in five years. Only three saw themselves leaving within the next two years. As for what they hoped to do after leaving the company, several hoped to work abroad, either as factory workers or as maids, others thought they would return to their home towns/villages to set up small businesses, and nearly all saw themselves starting families in their mid-twenties. None of the production operators saw themselves developing long-term careers in either the electronics sector or in a particular occupation associated with manufacturing. It is interesting to note that Graciela, the technician who has the best internal promotion prospects of the respondents, was the only one with a long-term career goal connected to her present employment. But she also saw herself leaving Storage Ltd. within two years. This is due in part to her wish to return to finish her engineering degree. But it also highlights the inverse relationship between the levels of education/technical training with attachment to the job. With more labor market options and higher expectations, workers like Graciela tended not to be as "committed" to their jobs as workers with fewer skills, less education, and fewer labor market options. This is similar to Integrated Production, discussed below, where production workers have a much higher level of education and higher aspirations, thus leading to a higher turnover rate despite a large number of positive incentives.

Thus workers' assessments of work at Storage Ltd. often seemed contradictory. Although many complained about the overtime and lack of rest days, they did find work at least tolerable and most had plans to stay with their current jobs for at least the next two years. Most mentioned the relatively high pay and were glad to be earning an income. For many, the job gave them their first taste of independent living away from watchful parents, increased leverage and status in their families, and control of their own wages. In addition, they enjoyed relatively high status as workers in a clean, high-tech multinational firm. Workers often referred to the work as "light" and "clean," particularly in contrast to the heavy, physical, and low-status work that awaited them in the provinces if they lost their factory jobs.

Understanding how workers situate work in the broader context of their lived experiences helps make sense of some of their seemingly contradictory statements. Work that might be quite demanding and intense can at the same time be relatively better than external labor market options and provide many nonmaterial benefits. These considerations, then, help explain workers' relatively high level of commitment—or at least attachment—to their jobs at Storage Ltd.

That said, by exploiting external inequalities, Storage Ltd.'s panoptic work regime creates a sense of worker autonomy, yet within the confines of organizational hierarchies, thus helping perpetuate unequal gender relations.

Individualized Responses at Integrated Production

Integrated Production's peripheral human resource (HR) approach deemphasizes strategic localization and focuses on internal positive incentives to boost worker commitment. However, in their hiring strategy, Integrated Production targets educated workers at all levels of its operation. Because the Philippines has an abundance of college graduates, the company has been able to build a workforce with impressive credentials that looked quite different than those at Storage Ltd. and Allied-Power. And reflecting the firm's other hiring criteria, it also has an older, more experienced, and more urban (but still primarily female) workforce. For example, although both Storage Ltd. and Integrated Production had been in full operation for three years, 90 percent of Storage Ltd.'s workers were below the age of twenty-four, while at Integrated Production, only 60 percent were below this age.

The firm's hiring preferences lead to an interesting dilemma. Their educated workers have both more labor market options, and much higher career and job expectations, making negative sanctions far less effective as a commitment strategy. In examining workers' responses to the firm and their feelings of ambivalence, we get a clearer picture of the complexity of commitment, the power of strategic localization, and the shortcomings of an individualized work regime that does not significantly share power and decision-making.

As described in chapter 3, workers' experiences of factory conditions at Integrated Production are often dualistic: tense and stressful in actual production, yet with a more laid-back atmosphere at the plant generally due to the "morale-building" strategies of the HR department. Agnes, a test operator, when asked how she liked her job made explicit reference to the stringent quality audits:

> [The job] is okay. But it's hard inside. You get in arguments with a lot of people. And you're not used to getting ITRs [inspection trouble reports]. Like for us, we work with units. If the units do not match the traveler [the card which follows the units through production and where various operators record their output], and the QA [quality assurance inspector] notices it, you get an ITR.

And if you get two ITRs, you get suspended for two weeks. If the QA sees something wrong, it gets right back to you.

Despite the pressure, workers can be motivated to work hard if direct control and monitoring is combined with workers' internalization of firm goals, which often is accomplished through behavioral training. For example, one operator commented:

> They give you a warning if you don't reach the quota, but it's impossible not to finish it because the managers are always asking, why this is so and that is so. It gets hard to joke with them. You have to be serious. It's just that you can't lose time, because every minute is valuable to Integrated Production company.
> *Interviewer: Is that your motto?*
> No, it's what's required, depending on what you do. For example, the machines I handle are expensive. I can't be losing time. Everything has to be recorded, otherwise, they ask, "Where did this minute go? Or those five minutes?" So everything you do is recorded. It's hard. In writing out the summary, I have to write down exactly how many minutes.

Another important reason workers may grant their discretionary effort is if they sense management treats them fairly, despite high demands. The firm's Employee Relations staff takes up the task of creating a sense of fairness among workers and has developed a multichanneled communications system to find out about problems quickly. For example, when asked how she might resolve a problem with a supervisor, one worker replied, "You can go straight to the HR. For example, an engineer cursed an operator, who then went to the HR. The HR talked to the engineer, then the engineer apologized to the operator."

Such responses to workers' issues has gone a long way in building a feeling of respect for the individual, regardless of the position. Yet at the same time, these communications forums do not fundamentally challenge management's power. The tension between keeping communication lines open while also enforcing management hierarchy is evident in Maria's assessment of the ER staff and it's activities:

> The [ER] meetings are for communication. It's for making the company better. It's good to have the upper management and workers talk so they look at us as equals. And at Integrated Production, they can say no one is higher, we're all equal. . . . They come to the line once a day, or if there's a problem. Like when Ms. L and Sir T came because we didn't want to go to the company outing because after the outing, the night shift had to report for work. We had to work, while for the day shift, the company shut down. So we on the night shift, we shouldn't have to work. So they came to explain this and that, why we need to join because it's a company affair and we all have to be together. . . . Although

we're treated as equals and we can talk with them, of course there's always respect, they're higher than us. And you can't be cursing them. But they are not
always scolding us. So it's okay.

In this instance, a potential collective act of resistance by workers was
avoided by quick intervention, "counseling" by the ER staff, and a belief by
workers that they had been listened to and treated fairly. But it is also important to note that this type of management-controlled communication,
while allowing for limited worker consultation, remains a far cry from grant-
ing operators substantive control over their work or active collective participation in decision-making that affects their jobs.

Nevertheless despite production demands, many workers felt they were
treated fairly and had a positive overall assessment of their workplace. When
asked if they were satisfied with their jobs and why, workers often responded
that they were satisfied, not with the wages per se, but particularly because
of the generous benefits. Regarding wages, one operator, when asked if her
wages of P5,700 per month were sufficient to support herself stated, "No,
they're too little. Way too little. See, I have both nothing and saved nothing
and it's been a year . . . they say [our pay] is low, but this is the first time I've
worked. But here, it's monthly [paid], not daily."

Despite her dissatisfaction with her salary, she was nevertheless satisfied
with the job overall, due mainly to the benefits:

Interviewer: So you like your job now?
Yes I do. After you've been here one year, they give you a computer, free. It's
worth US$1,200! . . . They can also send you to school. If you want to go to
school, you could. And if you have good grades, you get extra bonus.

As in the case of Storage Ltd., workers' backgrounds and external reference points also clearly play key roles in determining workers' level of commitment and attachment to their jobs. Many workers, particularly with
previous work experience, compared the firm's working conditions and benefits with those at other firms. Agnes, twenty, describes the factory job in a
warehouse she held before coming to Integrated Production:

I was there for two months. Wages are so low over there, it could drive you
crazy. I fainted, I didn't know [factory work] was like that. It was so hot . . .
and when the [break bell] rang, you got so happy and you didn't want it to ring
again, because it meant returning to work.
Interviewer: So you like it at Integrated Production?
Yes, here we're able to move around, go eat, compared to [the other firm],
where we had to remain seated and couldn't go anywhere. And in terms of benefits, [the other firm] seems oppressive.

Workers also measured their jobs in relation to its effect on their nonwork lives and their other work options. Similar to the young women workers at Storage Ltd., workers at Integrated Production found that their jobs provided prestige and a measure of independence from parents. Maria noted:

> Before, my Mom asked me not to go out just anywhere, or sleep over at a friend's house, or leaving without telling her when I'll be back. But now, it's okay with her, as long as I can take care of myself and have some money. . . . If I weren't working now, I'd still be in school or staying home, selling fish balls [laughter]. Or doing laundry or selling barbequed bananas.

Another worker, who compared her job at Integrated Production with those of her four siblings who all worked at various other electronics companies, explained, "Integrated Production is better. Because of the benefits, we have more benefits. . . . Integrated Production is even better than Intel because we get computers and they don't. My brother says, 'I have this, do you?' And I say, 'We have a bonus, do you?' My other siblings are jealous of me."

Thus, in part because of the benefits package, workers were relatively satisfied with and attached to their jobs, at least in the short term.

Yet the firm's recruitment strategy targeting college-educated women also meant higher expectations in terms of conditions and pay, and the desire to have a higher-status job than "just an operator." When asked about whether they were satisfied with their present jobs, many had answers similar to Joanne, a twenty-year-old who has been with the company as a production worker for a year: "No, not really, I don't want to be an operator all my life. I want to have a higher position . . . manager." While her ambitions were high, upon reflection, her actual promotion expectations were much more modest. She, like other operators interviewed, knows that a management or technical track is not practically open to her and that what she can realistically hope for is to become a trainer or to transfer to administration. While the pay in administration is no higher, the status is clearly superior. Joanne explained:

> I'd like to be a clerk in the office. It's much nicer. I'd like to vary my surroundings, you know, paper work.
> *Interviewer: Is the work there easier?*
> No, not really. But when you're in the office, they perceive you differently. If you're just a production operator, they can say that you're just wasting your education.
> *Interviewer: So are you embarrassed to be an operator?*
> Not really, I'm proud of my wages, they're high. And I'm proud that I am at Integrated Production.

The firm was quite aware of this potential dilemma and did take steps to provide career ladders to prevent turnover. Clara explained:

> I don't want to change jobs, it'd be a pity to lose this job, I'm happy here. I want to stay, but in a different position. HR. The super [supervisor] helps us achieve our goals. They don't want us to stagnate in our jobs, so they say when I get my computer, I can then try to get more education. I want to be in HR, communication. I don't like the business side. An HR director takes care of communication among the workers, that's what I want, HR director.

Yet the firm clearly cannot accommodate all the career ambitions of its operators, which may eventually have a negative affect on turnover. Workers too understood the limits of internal mobility for operators, and many did not see themselves at the firm even in two year's time. For example, Maria outlined how she saw her future:

> In two years? I don't know, if my friends leave, I wouldn't like it here. Maybe I'll be abroad. Anywhere, as long as the wages are high. . . . In five years, maybe I'll have a boy friend. Maybe I'll own my own home. . . . In ten years, I'll be married, if someone would accept me. And I'll stay home. No, I'll have a business. Selling accessories, like hair accessories. Truly, my ambition is to become a model. But I don't think it will happen, because I'm too short.

Another operator had this projection for her future at Integrated Production and beyond:

> I'll stay here, for now, it's kind of hard to be looking around [for work]. Maybe I can be a trainer, the salary is higher. My sister is a trainer. She just gives exams. If I lose this job, I'll look for something in semicon as well, it's better. . . .
> In two years, I'll be working and studying; I'd like to finish my studies before I try to get a better job. And I'd like to travel abroad. For example, Canada, there's a lot of offers. But I don't want to be a domestic, not that kind of job. I want to be with a company, as a psychologist maybe. In five years . . . I'll be twenty-three, still young. I'll have finished psychology, perhaps, then I'll have a new job, I'll be a boss myself. And I shall have helped a lot of people by then. In ten years . . . I'll be twenty-eight already. I'll be so old!! I'll have my own car.[4]

As is evident in their ambitious projections for themselves, workers at Integrated Production were clearly ambivalent about staying at the company, particularly if they could not move out of production. The firm's selective recruiting for highly educated workers seems to have led to a more professionally oriented, middle-class workforce with quite high expectations from work. The firm does try to compensate by providing a "fun" workplace, with

---

4. It is interesting to note in this case that she did not mention getting married at all.

open communication, behavioral training, and a generous slate of benefits. However, the firm ultimately stops short of offering those internal and positive incentives that *best* provide workers who have labor market options with real reasons to stay: high pay, internal career ladders, genuine decision-making power, and autonomy (Meyer and Allen 1997).

Ultimately, the lack of real upward mobility at Integrated Production—due in large part to the "masculinization" of higher-tech production and the higher, more specific requirements for better technical jobs—makes staying in regular production jobs unattractive to the less technically educated women workers. Like Storage Ltd. workers, these workers did value the high status of working for an elite company in an elite industry in the short-term. However, these workers also had both higher career ambitions and more labor market options than the Storage Ltd. workers and were thus less likely to stay committed to their jobs over the medium term.

It is no surprise then that if measured by turnover rates, Integrated Production's commitment strategies are the *least* effective of the three nondespotic firms studied, despite their HR approach and use of positive incentives. One disturbing conclusion is that from a managerial perspective, Storage Ltd.'s panoptic work regime and deep localization strategy seems to elicit greater worker attachment than Integrated Production's approach, which offers some positive perks and does not actively select for vulnerability.

Thus the ambivalent attitudes and higher turnover rates of Integrated Production workers seem to contradict the more optimistic theories of strategic HR and HR management that focus exclusively on the ability of management-defined positive incentives to elicit worker commitment (MacDuffie 1995). Yet despite the higher turnover, the firm's strategic HR approach might still be judged a "successful" labor control strategy by management. Through its behavioral training and individualized positive incentives, the firm was able to get workers to identify with firm goals and forestall any collective worker action.

An important question then remains: what kind of work regime might be able to build *genuine* worker commitment using positive rather than negative incentives? The case of Discrete Manufacturing, discussed below, suggests that a *comprehensive* and *bargained* set of positive incentives, which includes power sharing and collective negotiation of rewards, can lead to *both* a collectively empowered *and* committed workforce.

Inequality Transformed: Collective Gains at Discrete Manufacturing

Discrete Manufacturing shares several traits with the other two nondespotic multinational electronics firms in this book: it set up its factory just one year before the other two, has a workforce dominated by unskilled production operators, employs a workforce that is more than 75 percent women, and tries hard to build worker commitment to keep its high-tech production flexible yet stable. However, Discrete Manufacturing differs from the other firms in two crucial ways: it is unionized, and its workers are substantially older, with much longer job tenures. These two factors have substantially limited the scope of firm's localization and commitment strategies and also directly and indirectly affected the character of workers' experience at the firm.

Specifically, the case stands in contrast to the many studies of women workers in EPZs that emphasize only the exploitative side of electronics or similar factory work for women (Aldana 1989; Elson and Pearson 1984; Fernandez-Kelly 1983; Lim 1990). For while the firm initially hired these workers based on gendered stereotypes about women's "cheapness," "docility," and "appropriateness" for doing unskilled assembly work, their strong collective bargaining strength has helped these women defy many such stereotypes and has won them stable jobs, family-supporting wages, job mobility, and a long-list of negotiated benefits that the women value for helping increase their autonomy in the factory and within their own lives. The recursive relationship between working, collective organization, and the experiences of increased autonomy away from their jobs has also led in some cases to the opening of new areas of struggle at the workplace (Mills 2003). Thus their collective organization and the stability of their employment demonstrate that unionization and high-tech production are not only possible, but when combined, can realize at least some of the potential gains that industrialization and class-based organizing hold for all workers, men and women alike (Elson 1999; Hutchinson and Brown 2001).

Overall, workers interviewed had generally positive assessments of their jobs and relatively high levels of each type of commitment: effort, attachment, and loyalty.[5] But what is interesting are both the sources and the targets for worker commitment, which were often different than those suggested by other studies conducted primarily in industrialized settings (Lincoln and Kalleberg 1996). It is first important to review who the workers at Discrete Manufacturing are and their backgrounds. While 75 percent of the workers are women, their profile is nearly opposite of those at Storage Ltd. Their average age is twenty-nine years—versus twenty-one at Storage Ltd.—and less than 1 percent of the permanent workforce was below twenty-one (at Storage Ltd. it was a full 38 percent). Thirty-six percent of the workers at Discrete Manufacturing were also married—versus only 3 percent at Storage Ltd. In addition, because more than half the workforce had transferred from the old Manila plant, where getting hired only required

5. The author and research assistants interviewed a total of twenty-four workers. Interviews were conducted primarily at the plant and respondents were identified and chosen with the help of union officials.

a high school diploma, they had a much lower level of educational attainment than workers at Integrated Production and about the same as at Storage Ltd. These characteristics contribute to the workers' attitudes and behaviors toward management's work regime.

As discussed in chapter 3, management at Discrete Manufacturing had introduced a number of flexibility strategies and reorganization schemes— such as total quality management (TQM) and on- and off-line teamwork— in large part to discipline workers and increase worker effort. Workers' responses to these strategies turned primarily on how firm practices addressed their own interests, which were focused on internal issues of fairness, external issues of family life, and their complex notion of autonomy.

In terms of production and the pace of work, some found the shop floor pressures stressful. Carmen, who has a college degree in computer programming and came at eighteen from the distance province of Albay, explained:

> I was in Manila [at the old plant] for five months before transferring here. I'm not satisfied because the pay is low. Sometimes I get tired too. The work is hard. I'm assigned to glob top and I handle a machine. I also cover the alligns in the mid-section [of the line] and I handle another machine. Two machines I have to watch over and I'm standing the whole shift.

But management was also careful not to try and manipulate or micromanage all aspects of work or to even enforce all its own rules. For example, workers were free to take toilet breaks, and meal breaks remained untimed and often extended by workers, despite the official thirty-minute limit.

External references also had an impact on discretionary effort. Some workers felt that their jobs were so valuable that they needed to continually demonstrate their commitment in their work. But this commitment was also fostered by sense of negotiated fairness over pay and benefits. For example, Jaime, a twenty-three-year-old single operator from the distant province of Surigao del Norte, explained how he came to Discrete Manufacturing three years before and why he plans on staying:

> I worked before as one of the service crew at Tropical Hut—for five months, as a contractual. They didn't make anyone a regular [permanent employee] at Tropical. . . . Given the situation [in the labor market], I'm lucky to even be here, to be accepted at all! . . . I'm satisfied with work because there are lots of benefits and the salary is okay. We've been able to fight for them. Right now, I'm planning to stay because I'm already regular. I work hard and take care that I don't lose my job. It's tough if you're looking for work now. . . . If I lost my job, I would have to take whatever work's available. Because I really like it here, I'd probably look for something in semiconductors again. But I'll probably still be here at Discrete Manufacturing until I retire.

Other, particularly older, workers assessed their jobs by weighing previous experience with internal conditions. Francis, a twenty-eight-year-old married worker who has worked six previous jobs, most contractual and half in other manufacturing industries, said: "This job was a good opportunity because I have a family and here, there's job security. The work is okay too. I enjoy work. I have friends and management cooperates with me."

Despite generally positive job assessments, workers remained somewhat wary of management's commitment-building efforts attached to their work reorganization and flexibility schemes. For example, workers often questioned the limited returns to their contributions to such schemes. As Teresa, an operator active in the union stated,

> Some workers get excited about the quality program and the chance to represent the Philippines abroad in the [quality] competition. But then I ask them if they realize that all the benefits go to improving the profits of the company. We get one percent of the cost savings for our suggestions if they are implemented. And we're lucky to get even one percent. Mostly we just get token gifts, not money.

The disparity between management rhetoric and shop floor realities has limited the degree to which workers have internalized the goals and values of the firm. The less than enthusiastic response of workers to firm values does not, however, sever the link between autonomy and loyalty. Rather, workers did give their loyalty, but mainly to the union, which workers felt better articulated a broader, worker-defined sense of autonomy that reflects concerns both inside and outside the factory. At work, workers view autonomy as defined in opposition to management: having an independent source of power to extract concessions and protect themselves from arbitrary discipline. But possibly more important, workers also understand autonomy as increased power in and control over their nonwork lives. For many, particularly women workers, their jobs at Discrete Manufacturing have allowed them to secure wages high enough to either to delay marriage or to support their families. They have also been able to negotiate a large number of benefits that give them flexibility in and support for their lives outside of production.

From this point of view, worker autonomy has been best served not by management's benevolence, but by the union's negotiating prowess and workers' own collective strength to win worker-defined benefits at the bargaining table.

The Collective Bargaining Agreement

For workers, the collectively negotiated pay, benefits, and security of tenure clearly led to a sense of fairness and thus effort without direct disciplinary control by management. But the CBA and workers' voluntary effort did not necessarily lead to *organizational* commitment. By analyzing what workers bargained for in the CBA, it becomes clear that workers' lives outside of work remain of central importance to their negotiation of effort on the job and the creation of *job* commitment.

Highest on the workers' list of issues were pay and job security, particularly among older workers who transferred from the Manila plant and made up the core of the active union members. A majority of workers interviewed noted that in addition to supporting themselves and their own immediate families, they also remitted money home monthly to support other members of their family. As is typical in the Philippines, older siblings often pay the tuition and school fees for younger brothers and sisters. For example, Jaime, twenty-three and single, is the second oldest of seven siblings. He noted that he and his older sibling support the other five and that he remits P3,000 per month, or about half his wages back home. Similarly Shala, twenty-four and single, is the oldest of seven and sends home P2,500 a month to pay for the school fees of four siblings. Others send from P500 to P2,000 a month or whenever they could afford to. A number of workers also said they did not remit money home, but eight of eleven of these respondents were already married and most had children of their own.

In addition to wages, workers also negotiated a number of nonperformance based, end-of-the-year bonuses to help out during the Christmas season, when many workers return to the provinces and buy presents. While a Christmas bonus is a typical practice in many companies in the Philippines, workers at Discrete Manufacturing have been able to get three different bonuses equal to two and a quarter month's pay.

Still, while wages and bonuses at Discrete Manufacturing were much higher than at most other companies, most workers acknowledged that their wages were not enough to live on. In fact, twenty-one of the twenty-four workers interviewed said they regularly needed to borrow money to make ends meet or support family. The need for loans to support their nonworking expenses, then, was quite high and the wide array of loan and subsidy programs they have successfully negotiated into their CBA reflected workers' financial needs and nonwork priorities: housing, education, medicine, childbirth, sickness, bereavement, and emergency funds had all been secured.

Finally, workers negotiated to get time off. Already, Discrete Manufacturing workers worked fewer days per month than employees at Storage Ltd. and other firms. But they were also able to negotiate the highest number of sick and vacation days I encountered in visits to some twenty other firms and a number of other leaves, such as maternity/paternity leave, birthday leave, bereavement leave, union leave, and emergency leave.

From the list of benefits, it is clear that workers have traded effort at work for increased time and resources outside. As Edward, a married worker with kids put it, "You can't pay attention to your family if you're always at work. It should be eight hours of work and after that, eight hours of rest." Workers, then, value their jobs and are committed to them in large part because it affords them a family wage and real support and respect for their lives away from the factory.

Attached to the Job, Committed to the Union

As detailed in chapter 4, job attachment was also clearly influenced by conditions outside the plant, particularly for older workers. Francisco, a thirty-year-old married operator who studied electronics for one and a half years but didn't graduate, noted, "I work here because I wanted to use my educational background. And I like it too, because of all the benefits you can get. At my age, it would be tough for me to get another job. If I lost my job, I want another one like this one, because the work is quite easy." In assessing his future, he remarked jokingly, "I'll stay here at Discrete Manufacturing, so I can at least pay off my creditors."

Others also had somewhat negative assessments of the job itself, but still planned to stay for other reasons. Carmen, who complained above about the increased work pace and having to work two machines, nevertheless intended to stay for the long term:

> I'm not planning to look for anything else. I figure I'll be here at Discrete Manufacturing for the next ten years. There's a CBA. I don't plan to study anymore, since I've already had the chance to study. My projection for Discrete Manufacturing is that it's really lifetime employment. . . . If I lose this job, I'll probably look for another one in the semiconductor industry, probably here in [this EPZ]. Or I'd just go back to the province.

Many of the workers commented that the presence of the union made staying attractive. Juanita, a twenty-nine-year-old single female operator with the firm for four years, said she chose Discrete Manufacturing "because it has a union and is also a multinational corporation in the top ten." When asked what her future plans were, she commented, "[In two years], I'll still be here and still with the union. [In five years], I'll be studying for my master's degree while still at Discrete Manufacturing. [In ten years], I'll be at Discrete Manufacturing still, have a family, come what may."

Shala had a similar response: "The work is okay. I've adjusted to it already, and it's a big help financially. Right now, I have no plan to leave. . . . If I lost my job, I'd have two options: either start my own business [poultry or farming] or become an overseas contract worker." Regarding her future, she replied, "[In two years], I'll have my own family and still be with the union. [In five years], I'll have a child and be organizing the union. [In ten years], I'll have a happy family and still be with the union." That both Juanita and Shala equated being at Discrete Manufacturing with being active union member is testament to how synonymous the job and the union were in their eyes.

Workers also clearly understood that the firm's acceptance of the union has come only through worker resistance and a position of strength. When asked about the firm's view of the union, Rene noted, "They're agreeable with the members because the leader [of the union] is tenacious." Roland also commented, "The union is strong, so they [management] grant the union's demands." The most common response was that the firm "respects the union," but that the management is in constant tension, "always fighting against it." Finally, Edward noted, "The union is like a plague for them."

As for workers' own view of the union, they showed a relatively high level of support. Ramon, who was not particularly satisfied with his actual job, nevertheless commented, "I have a lot of faith/trust in them [*tiwala*]. They're our defense and the one that regular employees lean on for support." Patricia also replied, "Their mission is good [*maganda ang hangarin*]. It's much better with a union." Others also connected their assessment with what they felt the union has been able to get them. Edward stated, "It [the union] is very good. They've have been able to win what people need, especially hospitalization [benefits], wage increases, and the rights of people." Aris noted, "They really fight if they find some offense [by management], from something small to something really big; they provide a lot of protection for workers." The union has also been able to live up to expectations. As Sandy commented, "Before I started here, I wanted to see what it was like [with a union] because my Uncle has told me that it could help workers. And, as I've experienced, the union has been a big help." Similarly, Dennis stated, "They provide protection at work and make workers' rights known. We need the union because the capitalists oppress the workers. Before, I didn't see it that way, but that's the way I've actually experienced it."

Not all members were so enthusiastic about the union, particularly when it came to some of the union's more political work at a national level. But most agreed the union has played an important role in winning benefits. Ambeth, for example, has been with the company for seven years and has risen from a material handler to operator to trainer, and last year became a supervisor. He explained his problem with the union:

> I agree with union's goal to defend workers. They were good for getting wage increases and Lia was a very fair and responsible union president. But some officers and shop stewards are not so good. . . . Before in [the old plant] we would have large union meeting to discuss union policies. Here, in [the new plant], we were asked to sign union statements that I didn't agree with and wasn't able to discuss. We had no more big meetings. If we did, it was about action at Malacañang [the presidential palace/residence in Manila] and more about national politics than Discrete Manufacturing stuff.

Yet, for others, the political activities as part of a larger labor center's militant actions were a draw for workers to get more involved in wider issues.

Lia, the local union president, was also a vice president of the national labor center that was very active in national politics, and members were oriented, screened, and trained in order to find a core of committed cadre to work on more political issues. Weng, for example, has worked for five years with Discrete Manufacturing but had recently become active in the political wing of a militant regional union organization. When asked how and why she had become more politically active, she replied she wanted to do something more morally meaningful, rather than just working. She said trade unions handle the economic issues, but not directly the political and moral ones. Therefore, she began getting more active in the union's political wing. Indeed, the union at Discrete Manufacturing has proven to be a backbone for the larger labor center, which has recently suffered declining membership due to downsizing and increasing movements of other manufacturing offshore, competition from other unions, and active resistance from management at other firms.

While the union has been able to garner strong loyalties among workers, it is not always or necessarily at the total expense of organizational commitment. Indeed, in some cases, union members, and particularly some union leaders who have voluntarily taken up leadership positions in teams, have been able to combine their union and work roles, which has led to a high assessment of their overall jobs and a strong commitment to both firm and union. For example, Jose, a twenty-nine-year old married worker who has been with the company for five years, stated, "The job's quite good. I'm a union officer and have been the president of self-directed teams for four years. The benefits are really good, the company is strong, and there's a union." Workers such as Jose represent the epitome of negotiated commitment and the high potential for the collectively negotiated work regime.

The union has also been able to help build career ladders within the firm, giving workers access to technical and even managerial training. Yet one of the most surprising results from worker interviews was the large number of workers who did *not* want to move up to either technical or nontechnical supervisory positions. Because the union has bargained for seniority-based pay and regular pay increases, many operators were quite content to remain at Discrete Manufacturing as production operators. While there are many different production areas and requirements at different departments and stations, there is only one official job description for operator and only one pay scale. When asked how far he thought he could rise in the company, Roland, a twenty-eight-year-old operator who studied computer programming in college and has been with the company for five years, replied, "No idea. I'm content with where they've put me. I'm happy with where I am." Similarly, Janice, a twenty-nine-year-old college graduate in business management and operator for five years, said, "I'm not aiming for a high position, I don't want to be a supervisor. I enjoy my work." Finally, Patricia, a twenty-two-year-old operator who is also a licensed midwife, stated, "I really just want to stay an operator because the work is easier."

Indeed, even some who have risen from operators to supervisors have regrets. In a discussion, Ramon, who became a supervisor after many years as an operator, said it was particularly tough when CBA negotiations are going on, and only top management and the rank and file are involved. He said he and other supervisors "feel bad" because they see the wage increases and benefits package that the union is able to negotiate, while "we [supervisors] get the crumbs." He also noted that supervisors would like to get the same, but are not collectively represented. He said there had once been a rumor about forming a supervisors union, but that it never got past the rumor stage because of fear of retaliation or termination. He noted cautiously, "If there were to be a supervisors organizing drive, maybe we'd be looking for new jobs right now."

Gains for Women, Gains for Workers

As a unionized workforce at a leading multinational firm, workers at Discrete Manufacturing have been able to carve out an employment situation advantageous to themselves as workers and as women. Recall that the workforce in 75 percent female, 36 percent of whom are married. Interestingly, of the 744 married women workers, 72 percent had spouses that were unemployed. This means that a large percentage of the female workforce, 26 percent, were sole breadwinners for their families.

But it is also interesting to note that despite the relatively high number of married workers, 65 percent of the primarily older, female workforce is unmarried. While several female managers joked that there are "a lot of *sultera* [old maids] in this company," this pattern may also reflect the fact that women workers have been able to delay marriage beyond the average age of most Filipino women (Chant and McIlwaine 1995). Earning a decent wage has allowed workers some measure of independence (Wolf 1992). Shala, a twenty-four-year-old single worker, replied, "I had the chance to get married already, but when I started working, my boyfriend disappeared." Carmen, a twenty-two-year-old single worker, similarly replied, "If I wasn't working, I'd be back in the province . . . married, just sitting at home."

A number of the benefits also reflect women workers' priorities and attention to family issues. These include both maternity leave and childbirth subsidies that take into account the kind of delivery—caesarian, normal, or miscarriage—and thus the toll on the mother. The housing loan, education loan and subsidy, family medical and dental insurance, and bereavement assistance all help to directly finance family expenses beyond those of the individual worker. The large number of leaves also reflects women workers' priorities, as women in the Philippines are usually responsible for attending to family emergencies such as sick children or parents. Finally, the low and/ or interest-free loans have proven one of the most popular benefits as most workers need, at times, to borrow money.

That so many women at Discrete Manufacturing remained at their jobs and in the workforce after starting their own families and considered themselves "lifetime" employees is in stark contrast to other studies that have focused on the temporary nature of women's factory work (Fernandez-Kelly 1983; Ong 1987). In large part because the women had won collective power and some autonomy at work, they could also extend their understanding of autonomy to include increased power over their own lives. Indeed, Discrete Manufacturing offers an example of a work regime that seems to most fully develop the potential that export manufacturing holds for women workers (Hutchinson 1992). Ironically, Discrete Manufacturing originally recruited women with only high school educations as "cheap assembly hands" for work in the older, more labor-intensive plant. But this entry-level work for young, inexperienced women has contradictory effects. It can be both exploitative (in terms of maximizing surplus value by minimizing labor costs) as well as empowering (in terms of women gaining some control over and agency within their lives). But crucially, the unionization of Discrete Manufacturing allowed women workers a collective and institutional means of leverage vis-à-vis the company. Thus the arena for worker's agency is brought back into the workplace. No longer must workers be satisfied with simply earning "a" wage but can, through collective bargaining, focus on improving the character and quality of the entire employment relation.

Conclusion: The Contradictory Meanings of High-Tech Work

As the four cases demonstrate, the impact and meanings of work for labor in the advanced electronics sector are by no means unambiguous: it is impossible to speak simply of *either* exploitation *or* empowerment. Rather, this chapter has focused on the quite varied individual and collective understandings and responses that workers had to labor force participation and their employers' work regimes. So how can we assess such variation and what might it mean for electronics workers, the overwhelming majority of whom are women? One useful approach is to examine worker responses in light of what Chow and Lyter (2002, citing Molyneux 1985) have referred to as *practical* versus *strategic* gender needs and interests. These terms "differentiate immediate needs for daily provisions from the long-term needs of women as a social group to tackle the structural roots of unequal access to resources and control" (Chow and Lyter 2002). By expanding this lens to examine the practical and strategic interests of both women *and* workers, we can better assess the various outcomes and actions across the four cases.

The case of women workers at Allied Ltd. shows that collective action and unionization is possible, even among the most vulnerable workers and against relatively mobile employers in collusion with local officials. Allied workers, recruited for their vulnerability, initially recognized the practical benefits of their labor force participation. However, in response to the increasingly despotic work regime and the drastic layoffs, workers began to move from individual issues and frustrations toward more collective action. Here, the workers benefited tremendously from the help of a well-organized, local labor center. The labor center's sophisticated and multipronged approach to organizing was based on its own form of strategic localization, which forged ties with local community groups as well as international solidarity organizations.

Despite their successes, the Allied workers and the provincial labor alliance face two key limits. First, their organizing successes sparked a backlash from employers, zone officials, and the local government. As one organizer lamented, "Things have gotten much worse. It's now a war on unions . . . they're using all the resources at their disposal to bust the unions." This has meant that the labor alliance has had to move beyond trying to motivate workers with appeals to their practical interests. The organizer continued,

> We need new organizing tactics. Before, we'd organize based on workers' specific issues only. The big problem was that we only had a minority of [union] members that really wanted the union. The majority just wanted the issue solved. But if that issue did get solved, then there was no more base to be unified. Now, we have a more long-term vision. Now we want to be unified on issues *and* on the organization.

Thus organizers have realized the need to move to more strategic class and gender issues, but were finding it difficult to convince low-wage workers, who are often more concerned with the practical, day-to-day need just to survive and earn a living.

Second, the fact that Allied and its affiliated companies were willing and able to simply shut down their factories and threaten to move if labor problems intensified demonstrates the structural limits to labor organizing at the low end of the industry. Organizers at the provincial labor alliance admitted that the greatest obstacles to organizing were the continuing power of firms to close then reopen as nonunionized firms, and a complicit state and zone bureaucracy that allowed them to get away with it. Because Allied required such low capitalization, compared with giants like Storage Ltd. or Integrated Production, it was not difficult for them to simply shutter or move operations. This strategy had led to several hollow unionization "victories" for the provincial labor alliance at the Cavite EPZ, including at two of Allied's sister plants and several garment factories (WAC 2003b).

The limits of organizing at the low end of the market has led some labor groups to recognize the greater potential for unionization in higher-end firms, which have a lot more at stake and are more vulnerable to production disruptions. However, as we have seen in the cases of Storage Ltd. and Integrated Production, many of these higher-end firms are keenly aware of this very potential and undertook both antiunion localization strategies and internal education or union-substitution campaigns to prevent collective organizing.

Despite the "success" of Integrated Production and Storage Ltd. in forestalling collective worker action, it would be a mistake to overlook the many benefits their workers did draw from their employment. First, both cases confirm what a number of other studies have also pointed out: that multinational electronics firms tend to pay higher wages and provide better job security and working conditions than local factories or the informal sector (Joekes and Weston 1994; Rasiah 1995). While this can be a sign that non-union firms pay more to *avoid* unionization, at the individual level, workers clearly benefit (Deyo 1997). Second, although work is often quite demanding, workers themselves often had positive assessments of their jobs as it afforded them "clean" work and increased autonomy. As others have suggested, women benefit from factory employment generally as it can improve their bargaining power in their families and local communities (Elson 1999). Equally important for women at all levels of education was the higher status that work in the electronics industry provided. Even though women at Integrated Production were not necessarily proud of their status as production operators, they were quite proud to work in the electronics industry and particularly for Integrated Production. As the industry has become more high-tech, selective, and automated, the standing of workers in the industry has also improved, particularly vis-à-vis alternative employment opportunities in the informal economy, service sector, or in garment manufacturing. Freeman (2000, 61), in her study of low-paid women informatics workers in Barbados, notes a similar outcome: "Even though informatics does not present these women with a clear vehicle for actual class mobility, their work status gives them the sense that new identities are nonetheless possible."

Thus both Integrated Production and Storage Ltd. have in many ways helped their primarily female workforces to realize individual and practical gender interests through wage work. Nevertheless, there are clear limits to this sort of asymmetrical "empowerment" that leaves the hierarchical social relations within the firms intact. Analysis of the three case studies seems to suggest that the most potent vehicle for realizing the longer-term strategic class and gender interests would be through collective worker organization within firms that are larger, more permanent, and more able to negotiate real gains in exchange for stability and consent. The final case, Discrete Manufacturing, may provide just such an example of potential power realized: here a group of initially "marginal" women workers have in fact secured some very substantial benefits through industrial waged labor that give them increased income security, class mobility, and power over their own lives. But this potential has only been realized because of workers collective bargaining power, the willingness of Discrete Manufacturing's management to negotiate, and the ability of workers to extract concessions that reflect women workers' practical as well as strategic needs and interests.

Unfortunately, the successes at Discrete Manufacturing are not widely shared in the industry. Despite similar structural conditions of permanent employment across most of this industry, fewer than 10 percent of electronics firms have labor unions. It is also evident from the other two nondespotic cases that technological upgrading and the industry's general antiunionism are likely to reduce the chances of replicating the successes of workers at Discrete Manufacturing across the industry.[6]

6. On the industry's notorious antiunionism, see Aganon et al. 1997; Aldana 1989; Devinatz 1999; Hodson 1988; Hossfeld 1995; Milkman 1991; Pellow and Park 2002.

6 The Local and the Global

THE FOUR CASE studies presented in this book tell a multitude of stories of how firms organize high-tech production, how managers and state officials regulate localities and labor markets, and how workers experience and help constitute the complex confluence of work, power, and social inequality.

Taken together, the firms clearly demonstrate that it is possible to manufacture advanced electronics in peripheral and institutionally "thin" locations like the Philippines. The wide variation in their work regimes also shows that different types of work organization can achieve flexible production even within a single industry and in a single institutional context. But how representative are the four work regimes of the wider industry and will they continue to be viable?

As the case of Allied-Power shows, the use of direct coercion in production can lead to instability. Despotic control may unwittingly fan *individual* frustrations into larger *collective* grievances even among vulnerable and dependent workers, particularly if the political opportunity for organizing is available. While advanced manufacturers likely could not tolerate such uncertainty and instability, this does not mean that the "low-road" despotic work regime is necessarily unsustainable beyond the individual firm. In the case of Power Tech, for example, owners were quite willing to simply shut down a factory facing a union drive and erect another factory producing the same goods literally next door. This strategy was also common among garment factories operating in the same EPZ. This book has focused mainly on the advanced, higher end of the electronics market in claiming that commitment is essential to competitiveness. However, firms that choose *not* to compete on the cutting edge on the basis of quality and customization, but instead compete in more standardized, less-quality-sensitive markets may still find despotic work regimes a "viable" option, particularly given slack labor markets and political and institutional conditions that do not regulate closely conditions on the shop floor. As noted in chapter 2, although in terms of export value local MNC affiliates account for more than 80 percent of exports, there are a total of eight hundred electronics firms registered with PEZA and the Board in Investments, which means that in a large number of small producers and supplier firms—both local and MNC affiliates—more despotic systems still thrive. And while mortality rates for firms and businesses in this sector are high, forms of despotic work regimes seem to survive.

For each of the remaining three firms, pursuing competitive flexible production does require stability and the organization of worker commitment. Each firm, however, has chosen a different path. Storage Ltd., with its sophisticated product, the cutthroat character of its markets, and continued labor-intensiveness, pairs its neo-Taylorist labor process with external labor control strategies that leverage existing social, economic, and gender inequalities in the local labor market to secure attachment or what may be called alienative commitment. Storage Ltd.'s deep localization strategy, which depends on cooperation with local and national officials in the political construction and regulation of local labor market, clearly refutes the notion that foreign direct investment is somehow frictionless and that locations are completely substitutable.

Of the some twenty other electronics firms visited during the research period, Storage Ltd.'s panoptic work regime had one of the most comprehensive strategies to discipline and monitor its employees both inside and outside the factory. However, many elements of its labor control system—including electronic monitoring in production, strict enforcement of behavioral rules in the plant, its dispersed housing strategy, selective rural screening practices, and background checks—were also common in the industry. In fact Storage Ltd.'s tightly controlled conditions on the shop floor and preference for very young, inexperienced women with some (but not too much) college education were more typical of the Philippine industry as a whole than the other three firms presented. In addition, disk-drive assembly and testing is also one of the fastest growing segments in the Philippine industry. Two of the leading American and all five of the top Japanese firms in the disk drive industry have wholly owned plants in the Philippines, a confluence of major industry players that has been matched only recently by the surge in investments in China (McKendrick et al. 2000; *Economist* 2003a and 2003b).

Integrated Production also takes a technology- and engineer-led production strategy, but bundles its "hard" technology-led production with a soft, HRM strategy aimed at "purchasing" worker cooperation and commitment. Yet despite having a higher turnover rate than the other two flexible higher-tech firms, Integrated Production's peripheral HR work regime may prove to be one of the two most dominant work regimes as the Philippine electronics industry matures (the other being a version of the panoptic regime).

With its focus on internal strategies and positive incentives, the peripheral HR work regime offers a model that does not require the deep, sometimes illegal, and potentially more unstable localization strategies of the panoptic work regime. Instead, it can rely on the primarily preventative interventions and legal and disciplinary measures of the host state at the national and local levels to maintain the proper (antiunion) regulatory climate, which the firm seeks. The peripheral HR work regime, which is found in several other market-leading, image-sensitive, primarily American multinationals, also seems successful in generating enough—if not "high"—worker commitment for flexible production without disturbing management prerogative.

Finally, the collectively negotiated work regime at Discrete Manufacturing demonstrates that unionized competitive flexible production is indeed possible, if an unlikely choice. Both management and the union understand that flexibility is often as much about the balance of power and control over decision-making as about efficiency. Given the current institutional and regulatory conditions in the Philippines, it seems the collectively negotiated work regime is unlikely to spread widely through the industry. However, its demise will not be caused by its lack of competitiveness or viability, but because firms are equally concerned about defending and augmenting their control over production as they are about controlling their workforce.

Theoretical Reprise

The foregoing summary brings out both the distinctiveness of each case and their significant departures from the static image of EPZ firms as simply sites of "brutal coercion at the point of production" (Burawoy 1985, 265). In this section, I stress the cases' similarities, fitting them into a broader theoretical framework to address how global(izing) forces are locally constituted. Here, I return to and expand Burawoy's approach to the politics of production and reproduction to interject several critical points into the debates discussed in the introductory chapter.

Flexible Accumulation, High Performance, and the Labor Process

A way to revisit the theoretical debate over flexible or lean production is to ask, How new are the "new forms" of work organization apparent in the four case studies? In a nutshell, work organization seems to have advanced far beyond the despotic but clearly short of "high performance." On the first count, the case studies do show that coercion is not necessarily the default choice of all firms: firms may voluntarily chose a consent-based work regime even when operating in a developing country that lacks both the formal capacity to discipline large firms and a substantial welfare state that might raise workers' bargaining leverage with employers (Burawoy 1985). New and complex competitive demands in advanced high-tech manufacturing require a level of stability, quality, and flexibility that simple coercion does not seem to be able to deliver. The imperatives of fast speed to market, the greater integration of all portions of production, the rising costs of assembly and testing equipment, and the increasing importance of customers have all made assembly and testing a much more visible and important link in production. That assembly and testing has become the main point of interaction between the parent firm and the customer puts additional pressure on assembly and testing for flexibility and customization as well as upholding a corporate image of quality, competency, and control. Thus the flexibility and high-performance theorists are correct; flexibility and high quality (and I would add responsiveness to customers and speed to market) are keys to competitiveness. Three of the four cases demonstrate that these firms are flexible and do produce high-quality goods.

This does not mean, however, that the three more advanced firms in this book chose a "high-road" or a worker-centered high-performance solution to meet their needs for higher quality and stability. In contrast to arguments promoting a high-performance model, advanced electronics manufacturing does *not* seem to require radical shifts from traditional, top-down, hierarchical organization or genuine participation by broadly skilled and involved shop floor workers. Rather, productivity and quality gains at both Storage Ltd. and Integrated Production have been made mainly by marrying traditional scientific management tools, such as time-and-motion studies and statistical process control, with automation and real time manufacturing information systems. While more complex technologies do require the presence of technical skill on the shop floor, technicians and engineers fill this niche, not production operators. Despite much rhetoric of team production and greater worker autonomy, the cases presented here demonstrate that new manufacturing processes, such as computerized statistical process control, "best practice" procedures, and strict ISO quality certification guidelines are reducing the scope for worker input and further standardizing production processes at the level of the operator.

These cases confirm the findings of other studies of both semiconductor and hard disk drive manufacturers that many leading firms have chosen to meet the new competitive demands for fast production ramp up, strict quality control, and uninterrupted production with a strategy based on stable front-line workers, cheap technical and engineering labor, and the expansion of management prerogative (McKendrick et al. 2000; Chun 2001). Such results are also not confined to assembly manufacturing. For example, Appleyard and Brown (2001) found similar results in silicon wafer fabrication, the arguable core process of the semi-conductor industry. Finally, these results have also been found in other cases of advanced manufacturing in Southeast Asia, suggesting that the character of flexible accumulation presented here may be more widespread. Deyo (1997) finds competitive flexible production in the Thai auto industry, but is quick to distinguish it from the dynamic flexibility in more developed countries, such as in Northeast Asia, Europe, and North America—where competitive innovation has been driven from the shop floor, with mainly positive impact on worker skills and empowerment. In contrast, Deyo argues that the dynamic flexibility in developing countries in general may be less innovative and more "learning-based industrialization," which "relies on local adaptation of technologies and products developed elsewhere." As a result, and in line with the findings in this book, he argues, "employers may seek to institute forms of flexibility which minimize worker participation in favor of unchallenged managerial control over production" (Deyo 1997, 217). Competitive demands in global manufacturing, then, seem to be driving firms to develop more comprehensive strategies of production control to confront and contain these contradictory market pressures.

But at a lower level of analysis, the variation across the three case studies and firm actions in other domains outside the factory also demonstrate that the expansion of management prerogative does not *necessarily* lead to an inexorable push toward a single, new kind of work regime. Rather, work organization and labor control strategies still vary across firms, depending, in large part, on the character of a firm's product, the bargaining power of its workforce, and its parent firms' accumulation strategies. Crucially, the informal and formal regulation of the local labor market and the character of the export processing zones help create the external conditions that make such wide variation and the pursuit such divergent strategies possible.

Therefore, rather than reflecting uniformly "despotism," "hegemony," or even "hegemonic despotism," the organization of work in the Philippine electronics industry is best characterized as a flexible accumulation regime (Burawoy 1985). This broader approach encompasses not only how production, the labor process, and interfirm relations have be reorganized, but also how it is socially and political reproduced and regulated through other institutions, such as the national and local government, formal industrial relations, and the labor market (Gottfried 2000). Thus a key strength of the flexible accumulation approach is that it pays particular attention to political, social, and cultural factors that have often been treated as exogenous to production. In this book I have focused on the political apparatuses of flexible accumulation, or how firms actively engage the locally specific domains of reproduction to secure the stability and labor control needed under the more complex demands of advanced manufacturing.

Taking into account these external *political* interventions also helps explain the limits of even dynamic flexibility and why more advanced production in the Philippines may not led to greater worker empowerment. As a comparison of Discrete Manufacturing with both Storage Ltd. and Integrated Production makes clear, the absence of countervailing forces—here in the shape of collective bargaining power and a union contract—provides capital a much freer hand in designing and reproducing particular work regimes of their choosing. Deyo (1997, 216) makes a similar political point: that in Southeast Asia generally, the weakness of organized labor movements has contributed to "autocratic forms of industrial flexibility" that are "overwhelmingly attentive to managerial agendas driven by competitive economic pressures, to the exclusion of the social agenda of workers and unions."

Strategic Localization and the Political Construction of Labor Markets

Part of the "managerial agenda" under more complex competitiveness is to contain the contradictions between highly flexible, cost-conscious production and the need to maintain stability and high quality. How do firms accomplish this? As we have seen in the case studies, the primary way is for firms to use space and place strategically, taking advantage of uneven development and preexisting differences across localities (Massey 1995). Most fundamentally, firms invest in particular places and strategically localize elements of their work regimes in order to lower production costs, and/or better secure labor control and worker commitment. In the advanced electronics industry, assembly manufacturers flocked to the Philippines in the mid 1990s primarily to tap its unique labor pool: a range of relatively inexpensive, well-educated, and English-speaking workers, both skilled and unskilled. No other country in the region, save India, has a workforce with such an advantage in terms of English language abilities and education levels. As Markusen (1996) argues, in an era of increased globalization, production locales still remain "sticky" despite—or maybe because of—the increased mobility of capital across increasingly "slippery space."

Yet even Markusen underestimates the importance of local differences and the need for firms to engage their locality, particularly in export processing zones (EPZs), or what she calls "satellite platforms" (Markusen 1996, 304). Makusen, like other economic geographers and economic sociologists, tends to focus on local "embeddedness" either in terms of generic national industrial policies or dense linkages with supplier firms (Dicken 2003; Granovetter and Swedberg 2001). However, firms with weak local supplier links and insulated in EPZs from national regulations are still "embedded" in that they are dependent on the locality for the bulk of their labor force, which remains the "stickiest" of production inputs. Yet because labor, unlike other production inputs, is *socially* organized and must be reproduced, issues of power, inequality, and labor control necessarily play a role. Nevertheless, the fact that firms *must* engage the locality and its population does not dictate *how* they interact with their surroundings or workforces.

The best way, then, to understand how location matters is to analyze the interplay between work regimes, labor market segmentation, and the ways local labor markets are politically constructed. Across all four cases, and in the Philippines generally, continued labor market disadvantages for young rural women mean that firms practicing dynamic flexibility can secure educated, English-speaking workers at comparatively bargain basement prices. Other recent accounts of flexible production also make similar connections between work restructuring and labor markets, showing that different reorganization strategies can result in greater labor market segmentation. For example, Kalleberg (2003) points out that U.S. employers, looking to exploit both numerical and functional flexibilities, have been segmenting their workforces between "organizational insiders" or those on permanent contracts who enjoy the benefits of functional flexibility, and "organization outsiders" or those hired as temporary workers who bear the brunt and insecurity of numerical flexibility. But like others analyzing flexible production in more developed countries, Kalleberg draws an overly dualistic picture, in part because he treats the labor market and conditions therein as exogenous and assumes that functional flexibility requires a "high performance work organization" and the use of exclusively positive incentives for workers. Like early dual labor market theorists, he assumes that "good" primary sector jobs will be filled by "primary" labor market groups, to whom employers will have to pay high or "primary" sector wages (Reich et al. 1973; Colclough and Tolbert 1992).

But as we see in the Philippine cases, electronics firms seem to be able to have it both ways: the stability of a functionally flexible workforce made up almost entirely of "organizational insiders," yet at a relatively low cost and without sacrificing workplace control. This feat is accomplished not only by exploiting *existing* differences in the labor market, but also strategically intervening on the labor supply side of the equation—through, for example, recruiting young educated women from remote rural areas—in order to create a workforce more willing to work under such conditions. In other words, both the existence and *possibility* of a segmented labor market allows firms to restructure work in numerous ways and to choose from a wider range of labor control regimes to secure overall worker commitment. Thus the power differentials caused by existing gender ideologies and deep labor market segmentation may allow employers to have their cake—high quality and commitment—and eat it, too—at a low cost.

Critically, such extrafactory interventions into the labor market cannot be accomplished alone and under all conditions: employers rely on both local and national state actors to provide overall stability and to regulate local labor market institutions and conditions on terms favorable to investors. First, at the national level, the state actively reduces worker bargaining power through the creation of Export Processing Zones (EPZs) that openly tout—and deliver—their guarantee of "industrial peace." Second, the national state is also complicit in its strategic *non*enforcement of constitutional labor laws that explicitly promote full employment, equal opportunity, security of tenure, humane working conditions, and even a living wage. The Philippine Labor Code—which codifies a full slate of workers' rights, including the rights to self-organization and collective bargaining, "peaceful concerted activity" and the right to strike, and antidiscrimination protection—is also routinely violated with impunity (Azucena 1997). Third, the national government has also decentralized certain labor regulatory functions, such as minimum wage setting, to the regional level to the disadvantage of both organized and unorganized workers, who enjoy more collective power at the national level. Fourth, at the provincial and municipal level, state actors have been encouraged to set up labor management committee programs to intervene in industrial disputes. However, these committees often include many nonlabor and nonmanagement members, such as representatives of the provincial, municipal, and *barangay*-level governments and the provincial police, which usually means that in the case of industrial disputes, labor representatives remain vastly outnumbered and outvoted. Fifth, and also operating at the local level, are the government's nationwide Public Employment Service Offices (PESO), a relatively new government-sponsored labor market intermediary that helps coordinate rural recruiting and acts as a kind of employment service linking rural workers and local officials with zone and nonzone employers. But PESO, as shown in chapter 4, is heavily influenced by the direct needs of the employers and the political needs of the local government, thus serving primarily as a way for firms to reach even deeper into rural areas in search of educated but inexpensive and dependent labor. Finally, the privatization of technical training has led to the rise of a number of technical schools that permit firms to directly intervene and shape the labor supply. Here, the firms are able to "donate" their own machinery, provide instruction, select and monitor "their" trainees, but without the obligation to hire. In this case, training institutes have helped firms, such Discrete Manufacturing's new integrated chip plant, to try to specifically train female technicians in a direct attempt to break the labor market bottleneck for technical workers, but in a way that might open up a cheaper stream of (female) labor.

Thus state-firm interventions—and conspicuous noninterventions—into labor market regulation at a variety of scales help create what seem to be "unregulated" markets and surplus labor conditions, which clearly contribute to the weak and circumscribed bargaining power of even relatively well-educated Filipino workers. As a number of other studies have also recently pointed out, similarly "competitive" labor market conditions in areas from Texas (Amberg 2004) to Chicago (Theodore 2003) to Guangzhou, China (Fan 2002), are often the product of *political* contests aimed explicitly at constructing labor markets on terms favorable to particular interests. This confluence of political conditions, state intervention, and employer interests in the labor market thus create the conditions that allow manufacturing firms to restructure work and labor demand, while at the same time influencing labor market institutions and even labor supply. This gives management enormous power and allows them to secure exactly what Rubery and Wilkinson (1994, 32) predict: "primary workers at secondary prices."

Manufacturing Commitment

The labor market regulation approach above is not only helpful in explaining the diversity of employer choices in terms of production, it can also help illuminate debates about worker commitment and its relation to the myriad forms of labor control used throughout this book. Like the literature on flexible production and dual labor markets, much of the literature on worker commitment is written from the perspective of advanced industrialized countries and therefore tends to assume that in "high performance" or "modern" organizations, employers use noncoercive, corporatist approaches and positive incentives to increase worker commitment, which in turn leads to attachment and lower turnover (Appelbaum et al. 2000; Mueller and Boyer 1994; Halaby and Weakleim 1989; Lincoln and Kalleberg 1990). However, these individual-focused, psychology-based notions are at pains to explain the relatively high level of commitment given the often tightly controlled working conditions and the absence of many of the supposedly "necessary" antecedents in the Philippine case studies. Clearly missing are accounts of the negative incentives and external influences on workers' attachment, effort, and loyalty. It is not that positive incentives are unimportant in the Philippine cases, but it is the lack of attention to workers' own subjectivities and reference points outside of production that renders the traditional approaches incomplete. Instead, I revive in this book the notion of alienative commitment—which takes into account workers' weak labor market bargaining leverage—in order to bring the issue of power back in to the discussion of worker commitment and worker-employer negotiations. Viewed through a lens of power and considering labor market vulnerability, Hirschman's (1970) exit, voice, and loyalty model seems to both severely underestimate the costs *and* overestimate the possibilities of both exit and voice (Collinson 1994). Rather, worker decision-making regarding consent and resistance on the shop floor needs to take into account both internal and external power resources. As Wells (1996, 178) notes on the plight of undocumented workers in the Californian strawberry industry: "More than any other factor, it is workers' consciousness of their limited options and dispensability in the labor market that elicits their commitment to strawberry jobs [and] their engagement in their own recruitment and control."

It is also important to recognize that although workers' ultimately decide whether to consent to a work regime, commitment is by no means uniform in either form or effect. The different types of commitment found in the three nondespotic studies clearly have variable impacts on workers' agency and bargaining strength vis-à-vis their employers.

In the case of Storage Ltd., management carefully and strategically hired those for whom work offered a positive route to agency, but an agency in their own lives that did not directly challenge management authority in production. Workers at Integrated Production were slightly better off. The firm's strategy to soften its technical control in production with an HR strategy aimed at purchasing worker commitment meant workers enjoyed positive incentives such as a "free" computer and a workplace culture with open communication. In many ways, the attempt to control worker behavior and maximize nonparticipatory commitment resembles the "fun and surveillance" strategy in call centers found by Kinnie, Hutchinson, and Purcell (2000, 971). But just as in the call centers, this top-down strategy clearly limits genuine worker empowerment by suppressing any *collective* worker voice. Finally, the bargained commitment at Discrete Manufacturing demonstrates that worker consent can be secured using primarily positive incentives that provide more autonomy to workers. But these strategies must be attentive to broader understandings of autonomy, which workers defined as independence *from* management at work, as well as greater control over their lives outside of production.

Together, the cases help lay bare the continued struggle over production control that is at the heart of securing worker commitment. Like Burawoy's notion of consent, the critical theory of commitment presented here recognizes that advanced manufacturing firms need to convince workers to voluntarily cooperate with management in order to maintain production stability and avoid disruptions. But my approach also extends the work of Burawoy and other scholars who continue to focus on the shop floor (Salzinger 2003; Vallas 2003). A critical theory of commitment must consider attachment, effort, *and* loyalty, analyzing each in terms of its connection to labor control strategies and the *r*eproduction of social relations, both at work and beyond.

One worrisome conclusion from the cases examined in this book is that a deep localization strategy that targets workers for their inexperience, limited education, and labor market vulnerability can be even more "effective" than a strategic HR approach in generating worker attachment and creating a workforce that generally does not—or does not have the luxury to—look past their everyday practical interest in simply surviving. In fact, workers' view of their high-tech employment as "clean," "modern," and superior to other available jobs likely contributes to the low level of unionization in the electronics industry and reticence of many workers to even consider collective action (Hossfeld 1995). That firms, with the help of other actors, can manufacture commitment using a host of control strategies and negative incentives means that scholars—and workers themselves—must keep a skeptical eye on commitment programs promising worker "empowerment."

Neoliberal Localization and Regulatory State

A legitimate question to ask at this point is, why host country governments would help underwrite the strategic localization and worker commitment strategies of foreign employers to the detriment of local labor? Here it is useful to recall the rise of neoliberalism, and the recent consolidation of the regulatory state in the Philippines. As discussed in chapter 2, the country experienced a dramatic economic implosion in the turbulent 1980s, leaving it with a debt overhang of some $26 billion. To keep needed funds flowing into the country, the Philippines was forced to follow the domestic policy prescriptions of international banking institutions, focused on deregulation and the rolling back of domestic protection of the economy. By the 1990s, the Philippine state was promoting the neoliberal agenda without much external prodding, aggressively deregulating and privatizing the economy, freeing trade and investment, and restructuring the banking system in order to lure investors and make the Philippines more "globally competitive."

But it would be a mistake to characterize the regulatory state as *de*regulated or noninterventionist. As Jayasuriya (2001, 110) writes: "Intervention in the market is designed not for the purposes of compensation but to institute and stabilize the market system. The constitution of the regulatory state proceeds from the recognition that the existence of a market system does not inevitably guarantee the existence of market order, which requires the constitution of a range of institutions to insulate the market from what is seen as the corrosive influence of political bargaining." In other words, the regulatory state, and the process of neoliberalization generally, represents an ideological as well as policy transition from an emphasis on democratic government to economic governance.

This transition to economic governance entails the simultaneous "roll back" of state ownership, state bureaucracies, and domestic protectionist policies, and the "roll out" of new institutions that actively promote the global competitiveness of subnational investment locales by delivering nonmarket forms of coordination and unique local assets to support more complex and flexible forms of capital accumulation (Brenner and Theodore 2002). A clear example of one such emergent change is the spatial and institutional reorganization of Philippine export processing zones.

Indeed the development of the zone program and the transformation of zones from public manufacturing areas to privatized high-tech enclaves simultaneously represent neoliberal "moments of destruction"—such as the dismantling of national developmental schemes and exposure to foreign investment—and neoliberal "moments of creation"—such as the new collaboration between local real estate capital and foreign zone developers to link the zones to global industries rather than local areas. Another central element of the regulatory state is the construction of institutional autonomy, or the insulation of key governance institutions from broader societal politics. Here, the national government created a new entity, the Philippine Economic Zone Authority (PEZA), giving it the power of authority and its own legal jurisdictions in all matters within the zones. PEZA was also granted independence from other state institutions with power of inspection, such as the Department of Labor and Employment, which may harbor more prolabor (or less proinvestor) sympathies. Finally, the regulatory state, through PEZA and local governments, also strategically intervenes into the economy by spatial reorganizing of the zones, disorganizing worker housing, facilitating recruitment in rural areas, and directly suppressing union activities (Brenner and Theodore 2002).

If globalization, as some have argued, is the "politics of instituting markets on a global scale," then the Philippine case represents a clear and detailed example of neoliberal localization, or how globalization as a *process* is actually constituted at the national and local levels (McMichael 2005). But ultimately, states like the Philippines—in a poor position to bargain with global capital but in a strong position vis-à-vis local workers—are faced with a developmental dilemma. Under neoliberal globalization, regulatory states like the Philippines are downloaded with the responsibility for reproduction and stability but do not gain the real power or "transformative capacity" to improve their position within the global economic system. The Philippine state's weak position means it must often grant concessions to investors and cannot impose even productive restrictions, such as local content laws or technology transfer schemes, which could help the Philippines deepen the character of its industrialization and climb up the value-added ladder. Rather, the Philippine state—forced to compete with other regulatory states for investments—is reduced to "selling" local conditions or proffering its uniquely skilled labor to the global market. Ultimately, this limited and shallow "development" strategy leaves the Philippines, and similar regulatory states, complicit in their own exploitation.

Gender, Technology, and the Defeminization of High-Tech Work

One of the clear conclusions of this research is that, in myriad ways, flexible accumulation depends on the exploitation of spatial, social, and economic difference and the reproduction of inequality. The continued sexual division of labor and the ongoing gendered processes of work and production represent two key ways that gender and flexible accumulation are mutually constituted. This book, then, can be read alongside the voluminous literature on women and production, particularly in light of the way women in developing countries have long served as the "ideal labor" for mobile manufacturers and electronics assemblers. As Salzinger (2003) and many others have pointed out, global producers seeking lower-cost labor have exploited socially defined gender differences to construct desirable workers and "cheapen" women's labor, reinforcing, if not creating from scratch, the associations between docility, subordination, secondary status, and women's work (Mills 2003; Wright 2001).

In this book, all the firms used a gendered recruiting strategy, drawing on and reinforcing the almost universal trope of cheap, nimble-fingered female workers. Some firms went even further, selecting women in rural families with traditional obligations to support parents and siblings as a way to increase dependence as well as extrafactory control. This penetration into communities and families—what Elson (1994) has called the reproductive economy—is an extension of neoliberal localization. In this case, given the neoliberal context that allows them great latitude to intervene in areas outside of production, firms are simultaneously able to exploit women's subordinate position in the reproductive economy—such as women's secondary status and domestic obligations—while at the same time claiming "nonresponsibility for reproduction of human life" by paying below subsistence wages, hiring only single women, and firing pregnant workers (Acker 2004, 25). Thus like employers' increased control over the terms of work through their labor market intervention, neoliberal localization also seems to increase global capital's ability to reproduce gender inequalities through their actions in the productive and reproductive economies.

Again, such conditions do not mean that women workers themselves have not benefited from their employment in the electronics sector. In fact, for many, their mere labor force participation has brought increased mobility, some control over wages, and increased autonomy from parents. Nevertheless, as discussed in chapter 5, it is crucial to distinguish between *practical* gender interests—focused on short-term and individual survival—and *strategic* gender interests—concerned more broadly with reversing women's unequal access to resources and power. And on this account, it seems that only the women workers at unionized Discrete Manufacturing, who have been able to realize some of their structural potential as full-time *workers,* have gained enough collective strength to make some headway in transforming gendered power relations.

In a book focusing on high-tech work and industrial restructuring, it is also important to address the gendering of new technologies and technical change. Here, electronics firms are faced with a dilemma: on the one hand, in their pursuit of low labor costs for both production and some technical workers, management has sought to "feminize" the work and hire or train women workers. However, on the other hand, the technology itself and more technical positions are considered "masculine" and the proper realm for male workers. As Wright (2001) found in studying television assembly plants on the U.S.-Mexico border, women workers are often considered the "ideal" electronics assembly workers, yet are also considered unfit, inflexible, and untrainable for high-tech flexible production. This tension is evident in the gendering of the newly created position of technical operator at both Integrated Production and Discrete Manufacturing. At Integrated Production, management chose not to challenge gendered stereotypes about technology and hired all men for the technical operator positions. Although the "new" work did not differ significantly from that performed by female operators, management and male workers tried to "masculinize" the jobs and define the work as requiring "stamina," "strength," and "technical expertise" in order to align the jobs with male expectations. At Discrete Manufacturing's newest plant, which assembled high-end integrated semiconductors, management sought a slightly different strategy. Here, the firm sought to "feminize" the male-dominated technician and technical operators positions by explicitly training women and associating the jobs more closely to the "female" character of assembly work in an attempt to keep costs down and expand the labor supply.

The different approaches to gendering the new technical positions demonstrate the socially and politically constructed meanings of technology itself. The effects of upgrading on women and gender relations depend not on the technology alone but also on the political context. For example, women workers at Discrete Manufacturing have over the years experienced tremendous technological change: from fully manual to semiautomated to automated assembly. But their collective organization has helped sustain their employment and allowed many to reap at least some of the advantages of industrial upgrading (Elson 1999; Aganon et al. 1997). Thus it is the gendered notions about technology, coupled with the existing power structure, that shapes both the gendering of new technologies and the impact of technological change on women. Unfortunately, while the gendering of work is always an ongoing, contested, and negotiated process, conditions in the labor market and low unionization rates mean that there are few checks on management's power to implement the new technology in ways it sees fit. And as noted in chapter 3, the industry remains extremely gender stratified, with few chances for existing women workers to move out of the operator position. In addition, the current trend in the global electronics industry is clearly toward greater automation, which entails the wider use of (primarily male) technical operators, and the simultaneous downsizing of (primarily female) assemblers. With this trend already playing out in South Korea (Mehra and Gammage 1999), Malaysia (Ng and Mohamad 1997), and Taiwan (Berlik 2000), the stage is set in the Philippines for a "defeminization" of the electronics workforce just as the jobs are beginning to require greater technical skills.

Indeed, the Philippine electronics industry association recently developed new promotional materials extolling the virtues of the "the Great Filipino Electronics Worker." (excerpted at the beginning of chapter 2). This paradigmatic worker, pictured prominently in the glossy booklet, is touted as "highly skilled," has "extensive capability," and makes "state-of-the-art technology products . . . for export to the world." And while these claims echoed similar materials produced since the 1970s to lure investment, what is truly startling about this new campaign is the gender of its model employee: the "Great Filipino Electronics Worker" it turns out, is now named "Diego" (SEIPI 2003).

Political Implications

I end with a few words on the implication the book has for Filipino workers and their allies. Flexible accumulation and its political apparatuses present a particularly difficult dilemma for Filipino labor. When I began this research, I was struck by what seem to me a great potential for developing country workers to organize collectively in one of the world's leading industries. The industry, after all, consists of high-profile multinationals, sensitive to disruption, that employ nearly all permanent workers. But I soon found that firms—and their accomplices in the Philippine state—recognize this potential far more clearly than workers, and accordingly take active steps to head off potential collective worker action both inside and outside their plants. In many ways, this option is available to firms because the Philippine regulatory state has worked hard to create the appropriate local conditions and insulate its implementing institutions—such as the PEZA— from public scrutiny and political challenge. Indeed, the reorganization and reregulation of the zone program reflect all the elements of what Peck and Tickell (2002, 37) have called an "emergent combination of neoliberalized economic management and authoritarian state forms."

On the one hand, this means that workers and their allies face a far more complex situation than sheer coercion on the shopfloor. But on the other hand, labor control is never without its fissures and instabilities and the increased need for global firms to intervene in local labor markets may create political openings for workers away from production. One strategy, as the case of Allied Ltd. workers and the Cavite Labor Alliance demonstrates, is to "jump scales" or act beyond the local level. But jumping scales is much more than simply a move upward, for example, to the international level. Rather, scale jumping is "a process of developing networks of associations that allow actors to shift between spaces of engagement" (Herod 2001, 43). These spaces of engagement can be at a variety of levels along a continuum, from the micro to the macro, the local to the global. Herod rightly points out that "the contingent nature of such spaces of engagement means that frequently actors may attempt to 'go local' to outmaneuver more globally organized opponents" (2001, 43).

One such strategy of engagement is to "bring the politics back in" by unmasking the depoliticized facade of key economic regulatory institutions, from local-level wage and productivity boards to global-level banks and trade organizations. Another key space of engagement remains national politics, which fundamentally influences the terms and rules that govern bargaining power at the factory level. This might include pressuring national governments to enforce current labor laws and internationally-recognized labor rights, improve training and education, provide better housing, and invest more heavily in improving worker communities.

While acting on multiple scales seems like a tall order, the labor center at the Cavite EPZ and their provincial labor alliance again provide emergent examples of this multi-level strategy, or what I have called, "labor's scalar fix." This community-based approach to labor organizing draws on the consolidation and resources of the local working class community to agitate for local unionization rights from below, while also enlisting the support of international labor solidarity networks to maintain pressure from above. This scale jumping or "boomerang" strategy allows local workers to expand their "spaces of engagement" and thus bring increased pressure to bear on their employers as well as state officials (Keck and Sikkink 1998). As a number of studies have documented, a community-based labor center approach—often linked with other concerned actors and solidarity networks—has had some successes in organizing among low-wage workers in both developing and developed countries (AMRC 1998; Clawson 2003; Hadiz 2001; Hutchison and Brown 2001; Milkman 2000; Yates 2003).

Alas, capital is never static, and in the perpetual "dance of conflict" between workers and employers, the battle will likely bring a host of new challenges and, it is hoped, new opportunities for justice.

Bibliography

Acker, Joan. 2004. "Gender, Capitalism, and Globalization." *Critical Sociology* 30(1): 17–41.

Adler, Paul. 1995. "'Democratic Taylorism': The Toyota Production System at NUMMI." In *Lean Work: Empowerment and Exploitation in the Global Auto Industry,* edited by S. Babson, 207–21. Detroit: Wayne State University Press.

Aganon, Marie, R. P. Ofreneo, R. del Rosario, M. S. P. Ballesteros, and R. E. Ofreneo. 1997. *Strategies to Empower Women Workers in the Philippines Economic Zones: A Research Report Submitted to FNV and CNV (Netherlands).* unpublished.

———. 1998. "Strategies to Empower Women Workers in the Philippine Economic Zones." *Philippine Journal of Labor and Industrial Relations* 18 (1/2): 106–59.

Aldana, Cornelia. 1989. *A Contract for Underdevelopment: Subcontracting for Multinationals in the Philippine Semiconductor and Garment Industries.* Manila: IBON Databank Philippines.

Amberg, Stephen. 2004. "Governing Labor in Modernizing Texas." *Social Science History* 28 (1): 145–88.

Amirahmadi, H., and W. Wu. 1995. "Export Processing Zones in Asia." *Asian Survey* 35 (9): 828–49.

Amsden, Alice H. 1989. *Asia's Next Giant: South Korea and Late Industrialization.* Oxford: Oxford University Press.

———. 1994. "Why Isn't the Whole World Experimenting with the East Asian Model to Develop?" *World Development* 22 (4): 627–33.

Angel, David P. 1994. *Restructuring for Innovation: The Remaking of the U.S. Semiconductor Industry.* New York: Guilford Press.

Anonuevo, Carlos, ed. 1994. *Filipino Workers' Dictionary.* Manila: Friedrich Ebert Stiftung.

Appelbaum, Eileen, and Rosemary Batt. 1994. *The New American Workplace: Transforming Work Systems in the United States.* Ithaca: Cornell University/ILR Press.

Appelbaum, Eileen, Thomas Bailey, Peter Berg, and Arne L. Kalleberg. 2000. *Manufacturing Advantage: Why High-Performance Work Systems Pay Off.* Ithaca: Cornell University/ILR Press.

Appleyard, Melissa, and Clair Brown. 2001. "The Influence of Employment Practices on Manufacturing Performance in the Semiconductor Industry." *Industrial Relations* 40 (3): 436–71.

Armstrong, Pat. 1996. "The Feminization of the Labour Force: Harmonization Down in a Global Economy." In *Rethinking Restructuring: Gender and Change in Canada,* edited by Isabella Bakker, 29–54. Toronto: University of Toronto Press.

Arthur, Jeffrey. 1994. "Effects of Human Resource Systems on Manufacturing Performance and Turnover." *Academy of Management Journal* 37 (3): 670–87.

Asia Monitor Resource Center (AMRC). 1998. *We in the Zone: Women Workers in Asia's Export Processing Zones.* Hong Kong: AMRC.

Austria, Dr. Myrna S. 1999. "Assessing the Competitiveness of the Philippine IT Industry." PIDS Discussion Paper. Makati: Philippine Institute for Development Studies (PIDS).

Azucena, C. A. 1997. *Everyone's Labor Code.* Manila: Rex Bookstore.

Babson, Steve, ed. 1995. *Lean Work: Power and Exploitation in the Global Auto Industry.* Detroit: Wayne State University Press.

——. 1999. "Ambiguous Mandate: Lean Production and Labor Relations in the United States." In *Confronting Change: Auto Labor and Lean Production in North America,* edited by Huberto Juarez Nunez and Steve Babson, 23–50. Detroit: Wayne State University.

Bacon, Nicolas. 2001. "High Involvement Work Systems and Job Insecurity in the International Iron and Steel Industry." *Canadian Journal of Administrative Sciences* 18 (1): 5–17.

Bailey, Thomas, and Annette Bernhardt. 1997. "In Search of the High Road in a Low-Wage Industry." *Politics and Society* 25 (2): 179–202.

Baud, Isa, and Ines Smyth, eds. 1997. *Searching for Security: Women's Responses to Economic Transformations.* London: Routledge.

Becker, Howard. 1960. "Notes on the Concept of Commitment." *American Journal of Sociology* 66: 32–40.

Beeson, Mark, and Richard Robison. 2000. "Introduction: Interpreting the Crisis." In *Politics and Markets in the Wake of the Asian Crisis,* edited by R. Robison, M. Beeson, K. Jayasuriya, and H. Kim, 3–24. London: Routledge.

Bélanger, J., A. Giles and G. Murray. 2002. 'Towards a New Production Model: Potentialities, Tensions and Contradictions.' In *Work and Employment Relations in the High-Performance Workplace,* edited by G. Murray, J. Bélanger, A. Giles, and P. A. Lapointe. London and New York: Continuum.

Bello, Walden. 2000. "The Philippines: The Making of a Neo-Liberal Tragedy." In *Politics and Markets in the Wake of the Asian Crisis,* edited by R. Robison, M. Beeson, K. Jayasuriya, and H. Kim, 238–58. London: Routledge.

Bello, Walden, David Kinley, and Elaine Elinson. 1982. *Development Debacle: The World Bank in the Philippines.* San Francisco: Institute for Food and Development Policy.

Berger, Suzanne, and Richard Dore, eds. 1996. *National Diversity and Global Capitalism.* Ithaca, N.Y.: Cornell University Press.

Berlik, Gunseli. 2000. "Mature Export-led Growth and Gender Wage Inequality in Taiwan." *Feminist Economics* 6 (3): 1–26.

Bielby, D. D., and W. T. Bielby. 1988. "She Works Hard for the Money: Household Responsibilities and the Allocation of Work Effort." *American Journal of Sociology* 91: 1031–59.

Blauner, Richard. 1964. *Alienation and Freedom.* Chicago: Chicago University Press.

Bluestone, Irving, and Barry Bluestone. 1992. *Negotiating the Future: A Labor Perspective on American Business.* New York: Basic Books.

Bohn, Roger E. 2000. "The Low-Profit Trap in Hard Disk Drives, and How to Get Out of It." *Insight* (March/April): 6–10.

Borrus, Michael, Dieter Ernst, and Stephan Haggard, eds. 2000. *Production Networks and the Industrial Integration of Asia: Rivalry or Riches.* New York: Routledge.

Boswell, Terry, and Dimitris Stevis. 1997. "Globalization and International Labor Organizing." *Work* and *Occupations* 24 (3): 288–309.

Braverman, Harry. 1974. *Labor and Monopoly Capitalism: The Degradation of Work in the Twentieth Century.* New York: Monthly Review Press.

Brecher, J., and T. Costello. 1998. *Global Village or Global Pillage: Economic Restructuring from the Bottom Up.* 2d ed. Cambridge, MA: South End Press.

Brenner, Neil. 1999. "Beyond State-Centrism? Space, Territoriality, and Geographical Scale in Globalization Studies." *Theory and Society* 28: 39–78.

——. 2000. "The Urban Question as a Scale Question: Reflections on Henri Lefebvre, Urban Theory, and the Politics of Scale." *International Journal of Urban* and *Regional Research* 24 (2): 361–79.

——. 2004. *New State Spaces: Urban Governance and the Rescaling of Statehood.* Oxford: Oxford University Press.

Brenner, N., and Theodore, N., eds. 2002: *Spaces of Neoliberalism.* Oxford: Blackwell.

Brown, Clair, and Ben Campbell. 1999. *Technological Change, Training and Job Tasks in a High Tech Industry.* Available at http://socrates.berkeley.edu/~iir/

——. 2001. "Technical Change, Wages, and Employment in Semiconductor Manufacturing." *Industrial* and *Labor Relations Review* 54 (2a): 450–65.

——. 2002. "The Impact of Technological Change on Work and Wages." *Industrial Relations* 41 (1): 1–33.

Burawoy, Michael. 1979. *Manufacturing Consent.* Berkeley: University of California Press.

——. 1985. *Politics of Production.* London: Verso.

Bureau of Export Trade Promotion. 1998. *Industry Profile: Electronics.* Industrial Manufacturing Division, Department of Trade and Industry.

Bureau of Labor and Employment Statistics (BLES). various years. LABSTAT Updates. Manila: BLES/DOLE.

Bureau of Labor Relations/Department of Labor and Employment (BLR/DOLE). 1999 (unpublished). "Draft: Labor Management Schemes and Workers' Benefits in the Electronics Industry." Manila: BLR/DOLE.

Canlas, Corinne. 1991. *Calabarzon Project: The Peasant's Scourge.* Quezon City: Philippine Peasant Institute.

Cappelli, Peter, and David Neumark. 2001. "Do 'High-Performance' Work Practices Improve Establishment-level Outcomes?" *Industrial* and *Labor Relations Review* 54 (4): 737–75.

Carrillo, Jorge. 1995. "Flexible Production in the Auto Sector: Industrial Reorganization at Ford-Mexico." *World Development* 23 (1): 87–101.

Cavite Economic Zone (CEZ)/Philippine Economic Zone Authority (PEZA). Various years. *Monthly Reports.*

Chant, Sylvia, and Cathy McIlwaine. 1995. *Women of a Lesser Cost: Female Labour, Foreign Exchange, and Philippine Development.* London: Pluto.

Chellam, Raju. 2004. "Hard Time for Disk Drive Makers?" *Business Times* (Singapore) 6 October.

Chhachhi, Amrita. 1999. "Gender, Flexibility, Skill, and Industrial Restructuring: The Electronics Industry in India." Working Paper Series No. 296. The Hague: Institute of Social Studies.

Chow, Esther Ngan-ling, and Ray-may Hsung. 2002. "Gendered Organizations, Embodiment, and Employment among Manufacturing Workers in Taiwan." In *Transforming Gender and Development in East Asia,* edited by Esther Ngan-ling Chow, 81–104. London: Routledge Press.

Chow, Esther Ngan-ling, and Deanna Lyter. 2002. "Studying Development with Gender

Perspectives: From Mainstream Theories to Alternative Frameworks." In *Transforming Gender and Development in East Asia,* edited by Esther Ngan-ling Chow, 25–60. London: Routledge Press.

Chu, Yin-Wah. 2002. "Women and Work in East Asia." In *Transforming Gender and Development in East Asia,* edited by Esther Ngan-ling Chow, 61–80. London: Routledge Press.

Chun, Jennifer. 2001. "Flexible Despotism: The Intensification of Insecurity and Uncertainty in the Lives of Silicon Valley's High-Tech Assembly Workers." In *The Critical Study of Work,* edited by Rick Baldoz, Charles Kroeber, and Philip Kraft, 127–54. Philadelphia: Temple University Press.

Clawson, D. 1980. *Bureaucracy and the Labor Process: The Transformation of U.S. Industry, 1860–1920.* New York: Monthly Review Press.

——. 2003. *The Next Upsurge: Labor and the New Social Movements.* Ithaca: Cornell University/ILR Press.

Cohen, Aaron. 1995. "An Examination of the Relationship between Work and Commitment." *Human Relations* 48 (3): 239–43.

Colclough, Glenna, and Charles M. Tolbert. 1992. *Work in the Fast Lane: Flexibility, Divisions of Labor, and Inequality in High-Tech Industries.* Albany: State University Press.

Collins, Jane. 2003. *Threads: Gender, Labor, and Power in the Global Apparel Industry.* Chicago: Chicago University Press.

Collinson, D. 1994. "Strategies of Resistance: Power, Knowledge, and Subjectivity in the Workplace." In *Resistance and Power in Organizations,* edited by J. Jermier, D. Knights, and W. R. Nord, 25–68. London: Routledge.

Coronel, Sheila. 1995. "The Killing Fields of Commerce." In *Boss: Five Case Studies of Local Politics in the Philippines,* edited by Jose F. Lacaba, 1–32. Manila: Philippine Center for Investigative Journalism and Institute for Popular Democracy.

Danford, Andy. 1998. "Work Organization Inside Japanese Firms in South Wales: A Break from Taylorism?" In *Workplaces of the Future,* edited by Paul Thomas and Chris Warhurst, 40–64. London: MacMillian Press.

Dasmarinas, Julius. 1994. "At the Crossroads of Industrialization: A Peasant Village Reacts to Project Calabarzon." *Kayo Tao* 13: 82–96.

De Dios, Emmanuel, and Paul Hutchcroft. 2003. "Political Economy." In *Philippine Economy: Development, Policies, and Challenges,* edited by Arsenio Basilican and Hal Hill, 45–76. Quezon City: Ateneo University Press.

Dejillas, Leopoldo. 1994. *Trade Union Behavior in the Philippines 1946–1990.* Manila: Ateneo De Manila University Press.

Delbridge, Rick. 1998. *Life on the Line in Contemporary Manufacturing: The Workplace Experience of Lean Production and the "Japanese" Model.* Oxford: Oxford University Press.

Department of Labor and Employment (DOLE). 1997. "Labor Flexibilization: An Emerging Mode of Work Arrangement for Women in an Era of Globalization." Manila: DOLE.

Department of Labor and Employment/Bureau of Labor and Employment Statistics (DOLE/BLES). various years. *Yearbook of Labor Statistics.* Manila: DOLE.

Department of Trade and Industry (DTI)/ Bureau of Export and Trade Promotion (BETP). various years. Available at http://tradelinephil.dti.gov.ph/betp/Electronics.

Devinatz, Victor. 1999. *High-Tech Betrayal: Working and Organizing on the Shop Floor.* East Lansing: Michigan State University Press.

Deyo, Frederic. 1989. *Beneath the Miracle: Labor Subordination in the New Asian Industrialism.* Berkeley: University of California Press.

——. 1997. "Labor and Industrial Restructuring in Southeast Asia." In *The Political*

Economy of South-East Asia: An Introduction, edited by Garry Rodan, Kevin Hewison, and Richard Robison, 205–24. New York: Oxford University Press.

Dicken, Peter. 2003. *Global Shift: Reshaping the Economic Map in the Twenty-first Century.* New York: Guilford Press.

Diokno, Ma. Teresa. 1989. "The Failure of EPZs in the Philippines." In *Transnational and Special Economic Zones: The Experience of China and Selected Asian Countries,* edited by Teresa Carino, 130–51. Manila: De La Salle University Press.

Disk/Trend. 1999. *1999 Disk/Trend Report Summary.* Available at http://www disktrend.com/newsrig.htm.

Dohse, K., U. Jurgens, and T. Malsch. 1985. "From Fordism to Toyatism? The Social Organization of the Labor Process in Japanese Industry." *Politics and Society* 14 (2): 115–46.

Doyo, Ma. Ceres. 2000. "Cavite Export Zone: Cheapskates' Paradise." *Philippine Daily Inquirer,* May 1.

Dunning, John H. 1993. "Globalization and the New Geography of Foreign Direct Investment." *Oxford Development Studies* 26 (1): 47–70.

Economist. 2003a. "When You Can't Transplant Plant." *Economist* 15 (February): 64.

Economist. 2003b. "Is the Wakening Giant a Monster?" *Economist* 15 (February): 63–65.

Edralin, Divina. 2001. "Assessing the Situation of Women Working in CALABARZON." Philippines APEC Studies Center Network (PASCN) Discussion Paper No. 2001–14. Makati City: PASCN Secretariat.

Edwards, Richard. 1979. *Contested Terrain: The Transformation of the Workplace in the Twentieth Century.* New York: Basic Books.

Elger, Tony, and Paul K. Edwards. 1999. "Introduction." In *The Global Economy, National States, and the Regulation of Labour,* edited by Edwards and Elger, 1–41. London: Mansell.

Elson, Diane. 1994. "Micro, Meso, Macro: Gender and Economic Analysis in the Context of Policy Reform." In *The Strategic Silence: Gender and Economic Policy,* edited by Isabella Bakker, 33–45. London: Zed.

——. 1999. "Labor Markets as Gendered Institutions: Equality, Efficiency, and Empowerment Issues." *World Development* 27 (3): 611–27.

Elson, D., and R. Pearson. 1981. "Nimble Fingers Make Cheap Workers: An Analysis of Women's Employment in Third World Export Manufacturing." *Feminist Review* (spring): 144–66.

——. 1984. "The Subordination of Women and the Internationalization of Factory Production." In *Of Marriage and the Market,* edited by K. Young et al., 18–40. London: Routledge and Kegan Paul.

Ernst, Dieter. 1997. "From Partial to Systemic Globalization: International Production Networks in the Electronics Industry." BRIE Working Paper #98, Berkeley Roundtable on the International Economy, University of California at Berkeley.

——. 2002. "Digital Information Systems and Global Flagship Networks: How Mobile is Knowledge in the Global Network Economy." East West Center Working Papers, Economics Series No. 48. Honolulu: East West Center.

——. 2003. "Pathways to Innovation in Asia's Leading Electronics Exporting Countries: Drivers and Policy Implications" East-West Center Working Papers, Economics Series, No. 62. Honolulu: East West Center.

Ernst, Dieter, and David O'Connor. 1992. *Competing in the Electronics Industry: The Experience of Newly Industrialising Economies.* Paris: Development Centre of the OECD.

Esguerra, J., A. Balisacan, and N. Confessor. 2001. "The Philippines: Labor Market Trends and Government Interventions Following the East Asian Crisis." In *East Asian*

Labor Markets and the Economic Crisis: Impacts, Responses, and Lessons, edited by G. Betcherman and R. Islam, 195–244. Washington DC: World Bank and ILO.

Etzioni, A. 1975. *A Evaluation of Complex Organizations: On Power, Involvement, and their Correlates.* Rev. ed. New York: Free Press.

Evans, Peter. 1995. *Embedded Autonomy: States and Industrial Transformation.* Princeton, NJ: Princeton University Press.

——. 2000. "Fighting Marginalization with Transnational Networks: Counter Hegemonic Globalization." *Contemporary Sociology* 29 (1): 230–41.

Fan, Cindy. 2002. "The Elite, the Natives, and the Outsiders: Migration and Labor Market Segmentation in Urban China." *Annals of the Association of American Geographers* 92 (1): 103–24.

Felipe, Jesus. 2004. "Competitiveness, Income Distribution, and Growth in the Philippines: What Does the Long-run Evidence Show?" Manila: Asian Development Bank, Economics and Research Department Working Paper Series 53. Manila: ADB.

Fernandez-Kelly, M. 1983. *For We Are Sold, I and My People: Women and Industry in Mexico's Frontier.* Albany: State University of New York Press.

Ferner, Anthony. 1996. "Country of Origin Effects and HRM in Multinational Companies." *Human Resource Management Journal* 7 (1): 19–37.

Ferree, Myrna Marx. 1976. "Working-Class Jobs: Housework and Paid Work as Sources of Satisfaction." *Social Problems* 23: 431–41.

Foucault, Michel. 1977. *Discipline and Punish.* New York: Vintage Books.

Foulkes, Fred. 1980. *Personnel Policies in Large Nonunion Companies.* Englewood Cliffs, NJ: Prentice-Hall.

Fox, Julia. 2002. "Women Work and Resistance in the Global Economy." In *Labor and Capital in the Age of Globalization,* edited by Berch Berberoglu, 145–62. Lanham, MD: Rowan and Littlefield.

Freeman, C. 2000. *High Tech and High Heels in the Global Economy: Women, Work, and Pink-Collar Identities in the Caribbean.* Durham, NC: Duke University Press.

Freeman, Richard B., and Joel Rogers. 1999. *What Workers Want.* Ithaca: Cornell University/ILR Press.

Friedman, A. 1977. *Industry and Labour: Class Struggle at Work and Monopoly Capitalism.* London: Macmillan.

Frobel, F., J. Heinrich, and O. Kreye. 1978. *The New International Division of Labor: Structural Unemployment in Industrialised Countries and Industrialisation in Developing Countries.* Cambridge: Cambridge University Press.

Frundt, Henry. 1999. "Cross-Border Organizing in the Apparel Industry: Lessons from Central America and the Caribbean?" *Labor Studies Journal* 24 (1): 89–107.

Garrahan, P., and P. Steward. 1992. *The Nissan Enigma: Flexibility at Work in a Local Economy.* London: Mansell.

Godard, John, and John Delaney. 2000. "Reflections on the 'High Performance' Paradigm's Implications for Industrial Relations as a Field." *Industrial* and *Labor Relations Review* 53 (3): 482–503.

Gonzalez, Eduardo. 1999. "The Crisis of Governance in Asia: The Long Road Ahead for the Philippines." In *Reconsidering the East Asian Economic Model: What's Ahead for the Philippines?* edited by Eduardo Gonzalez, 207–23. Pasig City: Development Academy of the Philippines.

Gottfried, Heidi. 2000. "Compromising positions: emergent neo-Fordisms and embedded gender contracts." *British Journal of Sociology* 51 (2): 235–59.

——. 2004. "Gendering Globalization Discourses." *Critical Sociology* 30 (1): 9–15.

Graham, Laurie. 1995. *On the Line at Subaru-Isuzu.* Ithaca: Cornell University Press.

Granovetter, Mark, and Richard Swedberg, eds. 2001. *The Sociology of Economic Life*. Boulder, CO: Westview Press.

Granovetter, Mark, and Charles Tilly. 1988. "Inequality and the Labor Process." In *Handbook of Sociology*, edited by N. Smelser, 175–221. Newbury Park, CA: Sage.

Greider, William. 1997. *One World Ready or Not: The Manic Logic of Global Capitalism*. New York: Simon and Schuster.

Grint, Keith. 1998. *The Sociology of Work*. Cambridge: Polity Press.

Guest, David E. 1987. "Human Resource Management and Industrial Relations." *Journal of Management Studies* 24 (5): 503–21.

Hackman, J. Richard, and Ruth Wageman. 1995. "Total Quality Management: Empirical, Conceptual, and Practical Issues." *Administrative Science Quarterly* 40 (2): 309–44.

Hadiz, Vedi. 2001. "New Organising Vehicles in Indonesia: Origins and Prospects." In *Organising Labour in Globalising Asia*, edited by J. Hutchison and A. Brown, 108–26. London: Routledge.

Halaby, Charles, and D. Weakliem. 1989. "Worker Control and Attachment to the Firm." *American Sociological Review* 55: 634–49.

Hall, Peter, and David Soskice, eds. 2001. *Varieties of Capitalism: The Institutional Foundations of Comparative Advantage*. Oxford: Oxford University Press.

Hanson, Susan, and Geraldine Pratt. 1995. *Gender, Work and Space*. New York: Routledge.

Harvey, David. 1989. *The Condition of Postmodernity*. Cambridge: Basil Blackwell.

———. 2001. *Spaces of Capital*. New York: Routledge.

Hawes, Gary. 1987. *The Philippines State and the Marcos Regime: The Politics of Export*. Ithaca: Cornell University Press.

Henderson, J. W. 1989. *Globalisation of High Technology Production: Society, Space, and Semiconductors in the Restructuring of the Modern World*. London: Routledge.

———. 1994. "Electronics Industries and the Developing World: Uneven Contributions and Uncertain Prospects." In *Capitalism and Development*, edited by L. Sklair, 258–88. London: Routledge.

Herod, Andrew. 2001. *Labor Geographies: Workers and the Landscapes of Capitalism*. New York: Guilford Press.

Herrin, Alejandro, and Ernesto Pernia. 2003. "Population, Human Resources, and Employment. In *Philippine Economy: Development, Policies, and Challenges*, edited by Arsenio Basilican and Hal Hill. Quezon City: Ateneo University Press: 283–310.

Herzberg, F. 1966. *Work and the Nature of Man*. New York: Staples Press.

Herzenberg, Stephen. 1996. "Regulatory Frameworks and Development in the North American Auto Industry." In *Social Reconstructions of the World Auto Industry*, edited by Fred Deyo, 261–94. London: MacMillan Press.

Herzenberg, S., J. Alic, and H. Wial. 1998. *New Rules for a New Economy: Employment and Opportunity in Postindustrial America*. Ithaca: Cornell University/ILR Press.

Hirschhorn, L. 1984. *Beyond Mechanization: Work and Technology in a Postindustrial Age*. Cambridge: MIT Press.

Hirschman, Albert O. 1970. *Exit, Voice, and Loyalty: Responses to Decline in Firms, Organizations, and States*. Cambridge: Harvard University Press.

Hirst, Paul, and Jonathan Zeitlin. 1991. "Flexible Specialization vs. Post Fordism: Theory, Evidence, and Policy Implications." *Economy and Society* 20 (1): 1–56.

Hodson, Randy. 1988. "Good Jobs and Bad Management: How New Problems Evoke Old Solutions in High-Tech Settings." In *Industries, Firms, and Jobs: Sociological and Economic Approaches*, edited by George Farkas and Paula England, 247–79. London: Plenum Press.

———. 1989. "Gender Differences in Job Satisfaction: Why Aren't Women More Satisfied." *Sociological Quarterly* 30: 385–99.

———. 1991. "The Active Worker." *Journal of Contemporary Ethnography* 20 (1): 47–79.

———. 1997. "Individual Voice on the Shop Floor: The Role of Unions." *Social Forces* 75 (4): 1183–1213.

Hodson, Randy, and Teresa Sullivan. 1985. "Totem or Tyrant? Monopoly, Regional, and Local Sector Effects on Worker Commitment." *Social Forces* 63 (3): 716–31.

Hollingsworth, J. R., and R. Boyer, eds. 1997. *Contemporary Capitalism: The Embeddness of Institutions.* Cambridge: Cambridge University Press.

Hossfeld, Karen. 1995. "Why Aren't High-Tech Workers Organized? Lessons in Gender, Race, and Nationality from Silicon Valley." In *Working People of California,* edited by Daniel Cornfield, 405–32. Berkeley: University of California Press.

Humphrey, John. 1995. "Industrial Reorganization in Developing Countries: From Models to Trajectories." *World Development* 23 (1): 149–62.

Humphrey, John, and Hubert Schmitz. 1996. "The Triple C Approach to Local Industrial Policy." *World Development* 24 (12): 1859–77.

Hutchcroft, Paul. 1998. *Booty Capitalism: The Politics of Banking in the Philippines.* Quezon City: Ateneo de Manila University Press.

Hutchinson, Jane. 1992. "Women in the Philippines Garments Export Industry." *Journal of Contemporary Asia* 22 (4): 471–89.

Hutchinson, Jane, and Andrew Brown, eds. 2001. *Organizing Labor in Globalizing Asia.* London: Routledge Press.

Ichniowski, Casey, T. A. Kochan, D. Levine, C. Olson, and G. Strauss. 1996. "What Works at Work: Overview and Assessment." *Industrial Relations* 35 (3): 56–74.

Institute for Labor Studies (ILS). 1997. *Efficacy of Selected Labor Market Reforms in Promoting Globalization with Equity: The Philippine Case.* Manila: Department of Labor and Employment.

[Integrated Production] Human Resources. n.d. "Labor Relations Training Handbook." Unpublished company materials.

International Confederation of Free Trade Unions (ICFTU). 1996. *Behind the Wire: Antiunion Repression in the Export Processing Zones.* Brussels: ICFTU.

———. 2003. *Philippines: A Union Stronghold in the Export Processing Zones.* Trade Union Briefing No. 5, August. Brussels: ICFTU.

Japan International Cooperation Agency (JICA). 1990. *Interim Report, Master Plan Study.* Tokyo: JICA.

Jayasuriya, Kanishka. 2000. "Authoritarian Liberalism, Governance, and the Emergence of the Regulatory State in Post-crisis East Asia." In *Politics and Markets in the Wake of the Asian Crisis,* edited by R. Robison, M. Beeson, K. Jayasuriya, and H. Kim, 315–30. London: Routledge.

———. 2001. "Globalization and the Changing Architecture of the State: The Regulatory State and the Politics of Negative Co-ordination." *Journal of European Public Policy* 8 (1): 101–23.

———. 2003. "Introduction: Governing the Asia Pacific—Beyond the 'New Regionalism.'" *Third World Quarterly* 24 (2): 199–215.

Jessop, Robert. 2002. *The Future of the Capitalist State.* Cambridge: Cambridge University Press.

Joekes, Susan, and Ann Weston. 1994. *Women and the New Trade Agenda.* New York: UNIFEM.

Jomo, K. S. 1997. *Southeast Asia's Misunderstood Miracle: Industrial Policy and Economic Development in Thailand, Malaysia, and Indonesia.* Boulder, CO: Westview Press.

———. 2000. "Comment: Crisis and the Developmental State." In *Politics and Markets in the Wake of the Asian Crisis,* edited by R. Robison, M. Beeson, K. Jayasuriya, and H. Kim, 25–33. London: Routledge.

Jonas, Andrew. 1996. "Local Labour Control Regimes: Uneven Development and the Social Regulation of Production." *Regional Studies* 30 (4): 323–38.

Kalleberg, Arne. 2001. "Farewell to Commitment? Changing Employment Relations and Labor Markets in the United States." *Contemporary Sociology* 30 (1): 9–12.

——. 2003. "Flexible Firms and Labor Market Segmentation: Effects of Workplace Restructuring on Jobs and Workers." *Work and Occupations* 30 (2): 154–75.

Kalleberg, A., and P. Marsden. 1995. "Organizational Commitment and Job Performance in the U.S. Labor Force." *Research in the Sociology of Work* 5: 235–57.

Kang, David. 2002. *Crony Capitalism: Corruption and Development in South Korea and the Philippines.* New York: Cambridge University Press.

Kaplinsky, Raphael. 1995. "Technique and System: The Spread of Japanese Management Techniques to Developing Countries." *World Development* 23 (1): 57–71.

Katz, Harry C., and Owen Darbishire. 2000. *Converging Divergences: Worldwide Change in Employment Systems.* Ithaca: Cornell University Press.

Keck, Margaret, and Kathryn Sikkink. 1998. *Activists beyond Borders.* Ithaca: Cornell University Press.

Kelly, Philip. 2001. "The Political Economy of Local Labor Control in the Philippines." *Economic Geography* 77: 1–22.

Kenney, Martin, and Richard Florida. 1993. *Beyond Mass Production: The Japanese System and Its Transfer to the United States.* New York: Oxford University Press.

Kenney, M., R. Goe, O. Contreras, J. Romero, and M. Bustos. 1998. "Learning Factories or Reproduction Factories? Labor-Management Relations in the Japanese Consumer Electronics Maquiladoras in Mexico." *Work and Occupations* 25 (3): 269–304.

Kinnie, Nick, Sue Hutchinson, and John Purcell. 2000. "'Fun and Surveillance': the Paradox of High-Commitment Management in Call Centres." *International Journal of Human Resource Management* 11 (5): 967–85.

Klein, Naomi. 2000. *No Logo.* London: HarperCollins/Flamingo.

Knauss, Jody. 1998. "Modular Mass Production: High Performance on the Low Road." *Politics and Society* 26 (2): 273–96.

Kochan, Thomas, Harry Katz, and Robert McKersie. 1986. *The Transformation of American Industrial Relations.* New York: Basic Books.

Kochan, Thomas A., Russell D. Lansbury, and John Paul MacDuffie, eds. 1997. *After Lean Production: Evolving Employment Practices in the World Automobile Industry.* Ithaca: Cornell University Press.

Kochan, Thomas, and Paul Osterman. 1994. *The Mutual Gains Enterprise.* Boston: Harvard Business School Press.

Koo, Hagen. 2001. *Korean Workers: The Culture and Politics of Class Formation.* Ithaca: Cornell University Press.

Korton, David. 1995. *When Corporations Rule the World.* San Francisco: Berrett-Koehler.

Kraft, Philip. 1999. "To Control and Inspire: U.S. Management in the Age of Computer Information Systems and Global Production." In *Rethinking the Labor Process,* edited by Mark Wardell, T. Steiger, and P. Meiksins, 17–36. Albany: State University of New York Press.

Kuruvilla, Sarosh. 1996. "Linkages Between Industrialization Strategies and Industrial Relations/Human Resource Policies: Singapore, Malaysia, the Philippines, and India." *Industrial and Labor Relations Review* 49 (4): 635–58.

Lambert, Rob. 1990. "Kilusang Mayo Uno and the Rise of Social Movement Unionism in the Philippines." *Labour and Industry* 3 (2/3): 258–80.

Lamberte, Mario, C. B. Cororaton, M. Guerrero, and A. C. Oberta. 1999. "Impacts of the Southeast Asian Financial Crisis on the Philippine Manufacturing Sector." Discussion Paper Series No. 99-09. Makati: Philippine Institute for Development Studies.

Lee, Ching Kwan. 1998. *Gender and the South China Miracle: Two Worlds of Factory Women.* Berkeley: University of California Press.

Legge, K. 1995. *Human Resource Management.* Basingstoke, UK: Macmillan.

Lim, Joseph. 1999. "Growth with Equity." In *Reconsidering the East Asian Economic Model: What's Ahead for the Philippines?* edited by Eduardo Gonzalez, 15–82. Pasig City: Development Academy of the Philippines.

Lim, Linda. 1990. "Women's Work in Export Factories: The Politics of a Cause." In *Persistent Inequalities,* edited by Irene Tinker, 101–19. New York: Oxford University Press.

Lin, Vivian. 1987. "Women Electronics Workers Southeast Asia: The Emergence of a Working Class." In *Global Restructuring and Territorial Development,* edited by J. Henderson and M. Castells, 112–33. London: Sage Publications.

Lincoln, James R., and Arne L. Kalleberg. 1990. *Culture, Control, and Commitment: A Study of Work Organization and Work Attitudes in the United States and Japan.* Cambridge: Cambridge University Press.

——. 1996. "Commitment, Quits, and Work Organization in Japanese and U.S. Plants." *Industrial and Labor Relations Review* 50 (1): 39–60.

Linden, Greg, Clair Brown, and Melissa Appleyard. 2001. "The Semiconductor Industry's Role in the New Net Order." Center for Work, Technology, and Society Working Paper. Institute of Industrial Relations, U.C. Berkeley. Available at http://iir.berkeley.edu/worktech/SloanGlobal.pdf.

Loscocco, Karyn A. 1990. "Reactions to Blue-Collar Work." *Work and Occupations* 17 (2): 152–78.

Luthje, Boy. 2004. Global Production Networks and Industrial Upgrading in China: The Case of Electronics Manufacturing. East West Center Working Papers, Economics Series No. 74. Honolulu: East West Center.

MacDuffie, John Paul. 1995. "Human Resource Bundles and Manufacturing Performance: Organizational Logic and Flexible Production Systems in the World Auto Industry." *Industrial and Labor Relations Review* 48 (2): 197–221.

MacDuffie, John Paul, and John F. Krafcik. 1992. "Integrating Technology and Human Resources for High Performance Manufacturing: Evidence from the International Auto Industry." In *Transforming Organizations,* edited by Thomas A. Kochan and Michael Useem, 209–26. New York: Oxford University Press.

MacGregor, D. 1984. "Theory X and Theory Y." In *Organizations Theory,* edited by D. Pugh, 279–82. Harmondsworth, UK: Penguin.

Mair, Andrew. 1997. "Strategic Localization: The Myth of the Postnational Enterprise." In *Spaces of Globalization,* edited by Kevin Cox, 64–88. New York: Guilford Press.

Markusen, Ann. 1996. "Sticky Places in Slippery Space: A Typology of Industrial Districts." *Economic Geography* 72: 293–313.

Marsden, Peter, and Arne Kalleberg. 1993. "Gender Differences in Organizational Commitment." *Work and Occupations* 20 (3): 368–91.

Massey, Doreen. 1995. *Spatial Divisions of Labor: Social Structures and the Geography of Production.* New York: Routledge.

Maurek, Jan. 1999. *Making Microchips: Policy, Globalization, and Economic Restructuring in the Semiconductor Industry.* Cambridge: MIT Press.

McAndrew, John P. 1994. *Urban Usurpation: From Friar Estates to Industrial Estates in a Philippine Hinterland.* Manila: Ateneo de Manila University Press.

McCoy, Alfred, ed. 1993. *An Anarchy of Families: State and Family in the Philippines.* Madison: University of Wisconsin Center for Southeast Asian Studies.

McCoy, Alfred, and Ed C. De Jesus, eds. 1982. *Philippine Social History: Global Trade and Local Transformation.* Quezon City: Ateneo de Manila University Press.

McKendrick, D., R. Doner, and S. Haggard. 2000. *From Silicon Valley to Singapore: Location and Competitive Advantage in the Hard Disk Drive Industry.* Stanford, CA: Stanford University Press.

McLoughlin, Ian. 1996. "Inside the Non-Union Firm." In *The New Workplace and Trade Unionism,* edited by Peter Ackers, Chris Smith, and Paul Smith, 301–23. London: Routledge.

McMichael, Philip. 2005. "Globalization." In *Handbook of Political Sociology,* edited by T. Janoski, R. Alford, A. Hicks, and M. Schwartz, 587–606. New York: Cambridge University Press.

McMillan, Margaret, Selina Pandolfi, and B. Lynn Salinger. 1999. "Promoting Foreign Direct Investment in Labor-Intensive, Manufacturing Exports in Developing Countries." United States Agency for International Development/Consulting Assistance on Economic Reform (CAER II) Discussion Paper No. 42.

Medium-Term Comprehensive Employment Plan (1999–2004). 1999. Department of Labor and Employment. Manila: DOLE.

Mehra, Rekha, and Sarah Gammage. 1999. "Trends, Countertrends, and Gaps in Women's Employment." *World Development* 27 (3): 533–50.

Meyer, J. P., and N. J. Allen. 1997. *Commitment in the Workplace: Theory, Research, and Application.* Thousand Oaks, CA: Sage Publications.

Middlebrook, Kevin. 1996. "The Politics of Industrial Restructuring: Transnational Firms' Search for Flexible Production in the Mexican Automobile Industry." In *Social Reconstructions of the World Auto Industry,* edited by Fred Deyo, 200–232. London: MacMillan Press.

Milkman, Ruth. 1991. *Japan's California Factories: Labor Relations and Economic Globalization.* Los Angeles: Institute of Industrial Relations, University of California, Los Angeles.

———. 1997. *Farewell to the Factory: Auto Workers in the Late Twentieth Century.* Berkeley: University of California Press.

Milkman, Ruth, ed. 2000. *Organizing Immigrants: The Challenge for Unions in Contemporary California.* Ithaca: Cornell University Press.

Milkman, Ruth, and Kim Voss. 2004. *Rebuilding Labor: Organizing and Organizers in the New Union Movement.* Ithaca: Cornell University/ILR Press.

Mills, Mary Beth. 2003. "Gender and Inequality in the Global Labor Force." *Annual Review of Anthropology* 32: 41–62.

Montgomery, D. 1979. *Workers' Control in America: Studies in the History of Work, Technology, and Labor Struggles.* Cambridge: Cambridge University Press.

Moody, Kim. 1997. *Workers in a Lean World. Unions in the International Economy.* London: Verso Press.

Mowday, R. T., Steers, R. M., and Porter, L. W. 1979. "The Measurement of Organizational Commitment." *Journal of Vocational Behavior* 14: 224–47.

Mueller, Charles, and Jean Wallace. 1992. "Employee Commitment," *Work and Occupations* 19 (3): 211–37.

Mueller, Charles, and Marcia Boyer. 1994. "Employee Attachment and Noncoercive Conditions at Work." *Work and Occupations* 21 (2): 179–21.

Nadvi, Khalid, and Hubert Schmitz. 1994. "Industrial Clusters in Less Developed Countries: Review of Experiences and Research Agenda." Discussion Paper 339, Institute of Development Studies, University of Sussex, England.

National Commission on the Role of the Filipino Women. 1985. *Women Workers in the Philippines.* Manila: National Commission on the Role of the Filipino Women.

National Statistics Coordination Board (NSCB, various years). *Annual National Accounts.* Manila: NSCB.

——. (various years). *Labor Force Statistics.* Manila: NSCB.

National Statistics Office (various years). *Annual Survey of Establishments.* Manila: NSO.

——. (various years). *Philippine Labor Market, Key Indicators.* Manila: NSO.

Ng, Cecilia, and Maznah Mohamad. 1997. "The Management of Technology and Women in Two Electronics Firms in Malaysia." *Gender, Technology, and Development* 1 (2): 177–203.

Nunez, Huberto Juarez, and Steve Babson, eds. 1999. *Confronting Change: Auto Labor and Lean Production in North America.* Detroit: Wayne State University.

Ohara, Ken. 1977. "Bataan Export Processing Zone: Its Development and Social Implications." In *Free Trade Zones and Industrializations of Asia,* edited by Ken Ohara. Tokyo: Pacific Asia Resource Center.

Ong, Aihwa. 1987. *Spirits of Resistance and Capitalist Discipline: Factory Women in Malaysia.* Albany: SUNY Press.

O'Riain, Sean. 2000. "The Flexible Developmental State: Globalization, Information Technology, and the 'Celtic Tiger.'" *Politics and Society* 28 (3): 3–37.

——. 2004. *The Politics of High-Tech Growth.* New York: Cambridge University Press.

Ortiz, Sutti. 2002. "Laboring in the Factories and the Fields." *Annual Review of Anthropology.* 31: 395–417.

Osterman, Paul. 1999. *Securing Prosperity,* Princeton, NJ: Princeton University Press.

——. 2000. "Work Reorganization in an Era of Restructuring: Trends in Diffusion and Effects on Employee Welfare." *Industrial and Labor Relations Review* 53 (2): 197–96.

Parrenas, R. S. 2001. *Servants of Globalization: Women, Migration,* and *Domestic Work.* Stanford, CA: Stanford University Press.

Parker, Mike, and Jane Slaughter. 1988. *Choosing Sides: Unions and the Team Concept.* Boston: South End Press.

Parrado, Emillo, and Rene Zenteno. 2001. "Economic Restructuring, Financial Crises, and Women's Work in Mexico." *Social Problems* 48 (4): 456–77.

Patterson, Alan. 1999. "The Backend Takes the Lead." *Electronics Business Asia* (March). Available at: http://www.eb-asia.com.

Payne, John. 1999. "Electronic and Semiconductor Industries." Presentation at the Semiconductor.Electronics.Business@SEIPI: An Electronics Business Seminar, Sponsored by the Semiconductor and Electronics Industry of the Philippines, Inc. 29 October 1999. Makati, the Philippines.

Peck, Jamie. 1996. *Workplace: The Social Regulation of Labor Markets.* New York: Guilford Press.

Peck, Jamie, and Adam Tickell. 2002. "Neoliberalizing Space." In *Spaces of Neoliberalism,* edited by N. Brenner and N. Theodore, 33–75. Oxford: Blackwell.

Pellow, David, and Lisa Sun-Hee Park. 2002. *The Silicon Valley of Dreams: Environmental Justice, Immigrant Workers, and the High-Tech Global Economy.* New York: New York University Press.

Pena, Devon G. 1997. *The Terror of the Machine: Technology, Work, Gender, and Ecology on the U.S.-Mexico Border.* Austin: CMAS Books.

Penley, Larry, and Sam Gould. 1988. "Etzioni's Model of Organizational Involvement: A Perspective for Understanding Commitment to Organizations." *Journal of Organizational Behavior* 9: 43–59.

Perez, Daisy. 1998. "The Situation of Filipino Women Workers in Export Processing Zones and Industrial Enclaves in the Philippines." In *We in the Zone: Women Workers in Asia's Export Processing Zones,* 98–134. Asia Monitor Resource Center, Hong Kong: AMRC.

Pfeffer, Jeffrey. 1994. *Competitive Advantage through People.* Cambridge: Harvard Business School Press.

Philippine Economic Zone Authority (PEZA). (various years). *Economic Indicators.* Manila: PEZA.

Pineda-Ofreneo, Rosalinda. 1991. *The Philippines: Debt and Poverty.* Oxford: Oxfam UK.

Piore, Michael J., and Charles F. Sabel. 1984. *The Second Industrial Divide: Possibilities for Prosperity.* New York: Basic Books.

Piven, Francis F., and Richard Cloward. 2000. "Power Repertoires and Globalization." *Politics and Society* 28 (3): 413–30.

Political and Economic Risk Consultancy, Ltd. 1997. *Asian Labor Ratings.* Hong Kong: PERC.

Ponte, Remulo. 2001. "Cops Disperse Cavite Women Strikers." *Philippine Daily Inquirer.* 11 October.

Porter, Michael E. 2000. "Location, Competition, and Economic Development: Local Clusters in a Global Economy." *Economic Development Quarterly* 14 (1): 15–35.

Province of Cavite. 1999. *Provincial Profile: Cavite.* Trece Martires City: Province of Cavite.

Purcell, John. 1999. "Best Practice and Best Fit: Chimera or Cul-de-Sac?" *Human Resource Management Journal* 9 (3): 26–41.

Ramos, Carlos. 1991. *CALABARZON Master Plan: Issue and Implications.* Manila: Ramon Magsaysay Award Foundation.

Rasiah, Rajah. 1995. "Labour and Industrialization in Malaysia." *Journal of Contemporary Asia* 25 (1): 73–92.

Reich, M., D. M. Gordon, and R. C. Edwards. 1973. "A Theory of Labor Market Segmentation." *American Economic Review* 63: 359–65.

Rivera, Temario. 1994. *Landlords and Capitalists: Class, Family and State in Philippine Manufacturing.* Quezon City: Center for Integrative and Development Studies and University of the Philippines Press.

Rocamorra, Joel. 1994. *Breaking Through: The Struggle within the Communist Party of the Philippines.* Pasig City, Philippines: Anvil Press.

Rodan, Garry, Kevin Hewison, and Richard Robison, eds. 1997. *The Political Economy of South-East Asia: An Introduction.* Oxford: Oxford University Press.

Roethlisberger, F., and W. Dickson. 1939. *Management and the Worker.* Cambridge: Harvard University Press.

Rubery, Jill. 1994. "Internal and External Labour Markets: Toward an Integrated Analysis." In *Employer Strategy and the Labour Market,* edited by Jill Rubery and Frank Wilkinson, 37–68. Oxford: Oxford University Press.

Rubery, Jill, and Frank Wilkinson. 1994. "Introduction." In *Employer Strategy and the Labour Market,* edited by Jill Rubery and Frank Wilkinson, 1–36. Oxford: Oxford University Press.

Rubin, Beth. 1995. "Flexible Accumulation: The Decline of the Contract and Social Transformation." *Research in Social Stratification and Mobility* 14: 297–323.

Safa, Helen. 1981. "Runaway Shops and Female Employment: The Search for Cheap Labour." *Signs: Journal of Women in Culture and Society* 7: 418–33.

Salzinger, Leslie. 1997. "From High Heels to Swathed Bodies: Gendered Meanings under Production in Mexico's Export-Processing Industry." *Feminist Studies* 23 (3): 549–74.

———. 2003. *Gender in Production: Making Workers in Mexico's Global Factories.* Berkeley: University of California Press.

Saxenian, Annalee. 1994. *Regional Advantage: Culture and Competition in Silicon Valley and Route 128*. Cambridge: Harvard University Press.

Scipes, Kim. 1996. *KMU: Building Genuine Trade Unionism in the Philippines, 1980–1994*. Quezon City: New Day Publishers.

Seguino, S. 2000. "Accounting for Gender in Asian Economic Growth." *Feminist Economics* 6 (3): 27–58.

Semiconductor and Electronics Industry of the Philippines, Inc. (SEIPI). 1999a. *The Philippine Electronics Industry: An Industry on the Move*. Alabang: SEIPI.

———. 1999b. *SEIPI Digest*, No. 324 (March). Alabang: SEIPI.

———. 2001. *The Philippine Electronics Industry: Driver of the Philippine Economy*. Available at http://www.siepi.org.ph.

———. 2003. *SEIPI Information Book*. Alabang: SEIPI.

Semiconductor Industry Association (SIA). 1998. "America's Semiconductor Industry: Turbocharging the U.S. Economy."

———. 2004. "STATS: World Market Sales and Shares—1994–2003." Available at http://www.sia-online.org/downloads/market_shares_94–present.pdf., accessed November 13, 2004.

———. 2005. "SIA Projects 6 Percent Growth for Global Semiconductor Sales." Available at: http://www.sia-online.org/pre_release.cfm?ID=365.

Sewell, Graham. 1998. "The Discipline of Teams: The Control of Team-Based Industrial Work Through Electronic and Peer Surveillance." *Administrative Science Quarterly* 43 (2): 397–429.

Shaiken, Harley. 1995. "Lean Production in a Mexican Context." In *Lean Work: Power and Exploitation in the Global Auto Industry*, edited by Steve Babson, 247–59. Detroit: Wayne State University Press.

Shaiken, Harley, and Harry Browne. 1991. "Japanese Work Organization in Mexico." In *Manufacturing across Borders and Oceans: Japan, the United States, and Mexico*, edited by Gabirela Szekely, 25–50. La Jolla: Center for US-Mexican Studies, UCSD. Monograph Series 36.

Shaiken, Harley, S. Lopez, and I. Mankita. 1997. "Two Routes to Team Production: Saturn and Chrysler Compared." *Industrial Relations* 36: 17–45.

Sidel, John T. 1998. "The Underside of Progress: Land, Labor, and Violence in Two Philippine Growth Zones, 1985–1995." *Bulletin of Concerned Asian Scholars* 30 (1): 3–12.

———. 1999. *Capital, Coercion, and Crime: Bossism in the Philippines*. Stanford, CA: Stanford University Press.

Sklair, Leslie 1989. *Assembling for Development: The Maquila Industry in Mexico and the United States*. London: Unwin Hyman.

Smith, Neil. 1990. *Uneven Development: Nature, Capital, and the Production of Space*. Basil: Blackwell.

Smith, Vicki. 1994. "Braverman's Legacy: The Labor Process Tradition at 20." *Work and Occupations* 21 (4): 403–22.

———. 1998. "The Fractured World of the Temporary Worker: Power, Participation, and Fragmentation in the Contemporary Workplace." *Social Problems* 45 (4): 411–30.

———. 2001. *Crossing the Great Divide: Worker Risk and Opportunity in the New Economy*. Ithaca: Cornell University/ILR Press.

Soskice, David. 1999. "Divergent Production Regimes: Coordinated and Uncoordinated Market Economies in the 1990s." In *Continuity and Change in Contemporary Capitalism*, edited by Herbert Kitschelt et al., 101–33. New York: Cambridge University Press.

Southall, Roger. 1988. *Trade Unions in the New Industrialization of the Third World.* London: Zed Books.

Spar, Debora. 1998. "Attracting High-Technology Investment: Intel's Costa Rican Plant." Foreign Investment Advisory Service Occasional Paper 11. Washington DC: World Bank.

Standing, Guy. 1992. "Identifying the 'Human Resource Enterprise': A South-East Asian Example." *International Labour Review* 131 (3): 281–90.

———. 1999. "Global Feminization through Flexible Labor: A Theme Revisited." *World Development* 27 (3): 583–602.

Stichter, Sharon, and Jane Parpart, eds. 1990. *Women, Employment, and the Family in the International Division of Labor.* Philadelphia: Temple University Press.

Storper, Michael, and Richard Walker. 1983. "The theory of labour and the theory of location." *International Journal of Urban and regional Research* 7: 1–41.

———. 1989. *The Capitalist Imperative: Territory, Technology and Industrial Growth.* Oxford: Basil Blackwell.

Strange, Susan. 1996. *The Retreat of the State.* Cambridge: Cambridge University Press.

Sturgeon, Timothy. 1997. "Does Manufacturing Still Matter? The Organizational Delinking of Production from Innovation." *Berkeley Roundtable on the International Economy.* Paper BRIEWP92B. Available at *http://repositories.cdlib.org/brie/BRIEWP92B.*

Swyngedouw, E. A. 1997. "Neither Global or Local: 'Glocalization' and the Politics of Scale." In *Spaces of Globalization,* edited by Kevin Cox, 137–66. New York: Guilford Press.

Taplin, Ian. 1995. "Flexible Production, Rigid Jobs: Lessons from the Clothing Industry." *Work* and *Occupations* 22 (4): 412–39.

Tarrow, Sidney. 1998. *Power in Motion: Social Movements and Contentious Politics.* Cambridge: Cambridge University Press.

Taylor, Frederick W. 1911. *The Principles of Scientific Management.* New York: Harper and Brothers.

Tecson, Gwendolyn. 1999. "The Hard Disk Drive Industry in the Philippines." The Information Storage Industry Center, Report 99–01, University of California, San Diego.

Theodore, Nik. 2003. "Political Economies of Day Labour: Regulation and Restructuring of Chicago's Contingent Labour Markets." *Urban Studies* 40 (9): 1811–1828.

Tiglao, Rigoberto. 1996. "Field of Dreams." *Far Eastern Economic Review* (April 4): 50–51.

Tilly, Chris, and Charles Tilly. 1994. "Capitalist Work and Labor Markets." In *The Handbook of Economic Sociology,* edited by Neil Smelser and Richard Swedberg, 283–312. Princeton, NJ: Princeton University Press and Russell Sage Foundation.

Torres, Carmela. 1993. "External Labor Flexibility: The Philippine Experience." *Philippine Journal of Labor and Industrial Relations* 15 (1/2): 97–130.

Tsutsui, William. 1998. *Manufacturing Ideology: Scientific Management in Twentieth-Century Japan.* Princeton, NJ: Princeton University Press.

Tzannatos, Z. 1999. "Women and Labor Market Changes in the Global Economy: Growth Helps, Inequalities Hurt, and Public Policies Matter." *World Development* 27 (3): 551–69.

Vallas, Steven P. 1999. "Rethinking Post-Fordism: The Meaning of Workplace Flexibility." *Sociological Quarterly* 17 (1): 68–101.

———. 2003. "Why Teamwork Really Fails: Obstacles to Workplace Change in Four Manufacturing Plants." *American Sociological Review* 68 (2): 223–50.

Vallas, Steven, and John Beck. 1996. "The Transformation of Work Revisited: The Limits of Flexibility in American Manufacturing." *Social Problems* 43 (3): 501–22.

Van Heerden, Auret. 1998. "Export-Processing Zones: The Cutting Edge of Globalization?" Paper prepared for Global Production and Local Jobs: New Perspectives on Enterprise Networks, Employment, and Local Development Policy, International Institute for Labour Studies, Geneva, May 9–10.

Vasquez, Noel. 1987. *Mobilizing Surplus Labour through International Exchange: Philippine EPZs, Overseas Employment, and Labour Subcontracting.* Manila: Brotherhood of Asian Trade Unionists and Ateneo Center for Social Policy.

Villegas, Edberto M. 1989. *The Political Economy of Philippine Labor Law.* Quezon City: Foundation for Nationalist Studies.

Yuen Kay, Chung. 1994. "Conflict and Compliance: Workplace Politics of a Disk Drive Factory in Singapore." In *Workplace Industrial Relations and the Global Challenge,* edited by J. Belanger, P. K. Edwards, and L. Haiven, 190–233. Ithaca: Cornell University/ILR Press.

Yun, Hing Ai. 1995. "Automation and New Work Patterns: Cases from Singapore's Electronics Industry." *Work, Employment, and Society* 9 (2): 309–27.

Wade, Robert. 1990. *Governing the Market: Economic Theory and the Role of the Government in East Asian Industrialization.* Princeton, NJ: Princeton University Press.

Wade, Robert, and F. Veneroso. 1998. "The Asian Crisis: The High Debt Model versus the Wall Street-Treasury-IMF Complex." *New Left Review* 228 (1): 3–24.

Weiss, Linda. 1998. *The Myth of the Powerless State.* Ithaca: Cornell University Press.

Weiss, Linda, and John Hobson. 2000. "State Power and Economic Strength Revisited: What's So Special about the Asian Crisis." In *Politics and Markets in the Wake of the Asian Crisis,* edited by R. Robison, M. Beeson, K. Jayasuriya, and H. Ki, 53–74. London: Routledge.

Wells, Miriam. 1996. *Strawberry Fields: Politics, Class, and Work in California Agriculture.* Ithaca: Cornell University Press.

West, Lois. 1997. *Militant Labor in the Philippines.* Philadelphia: Temple University Press.

Whitley, R. D. 1999. *Divergent Capitalisms: The Social Structuring of Change and Business Systems.* Oxford: Oxford University Press.

Windell, James, and Guy Standing. 1992. "External Labour Flexibility in Filippino Industry." Working Paper No. 59, Labor Market Analysis and Employment Planning, World Employment Programme Research. Geneva: International Labour Office.

Wolf, Diane Lauren. 1992. *Factory Daughters: Gender, Household Dynamics, and Rural Industrialization in Java.* Berkeley: University of California Press.

Womack, James P., Daniel T. Jones, and Daniel Roos. 1990. *The Machine that Changed the World.* New York: Harper Perennial.

Wong, Poh-Kam. 2001. "Flexible Production, High-Tech Commodities, and Public Policies: The Hard Disk Drive Industry in Singapore." In *Economic Governance and the Challenge of Flexibility in East Asia,* edited by F. Deyo, R. Doner, and E. Hershber, 191–216. Lanham, MD: Rowan and Littlefield.

Woo, Jung-en. 1991. *Race to the Swift: State and Finance in Korean Industrialization.* New York: Columbia University Press.

Workers' Assistance Center. 1996. *Ang Kakagayan ng mga Manggagawa sa Cavite Export Processing Zone* [The Condition of Workers in the Cavite Export Processing Zone]. Rosario: Workers' Assistance Center.

——. 1999. *Manggagawa Manlilikha.* Various issues.

——. 2001a. "Workers Assistance Center, Inc. Cavite Philippines" (organization's promotional brochure).

——. 2001b. [Ultra Electronics, Ltd.] Factsheet. Unpublished.

——. 2003a. "A Study on the Conditions of the Workers in the Electronics Industry in

Selected Economic Zones in Cavite." available at: http://daga.dhs.org/atnc/electronic/reports/reports.htm. accessed on 3 March 2005.

——. 2003b. "A preliminary study on the situation and the gender struggle of the women workers in selected factories at the Cavite Economic Zone" available at: http://daga.dhs.org/atnc/archives/dossiers/dossiers.htm. accessed on 3 March 2005.

World Bank. 1993. *The East Asian Miracle: Economic Growth and Public Policy.* Oxford: Oxford University Press.

——. 1997. "Philippines: Managing Global Integration, Volume II." Report No. 17024-PH, Background Papers, Poverty Reduction, and Economic Management Sector, East Asia and the Pacific Office. Washington, DC: World Bank.

——. 1999. "Philippines: The Challenge of Economic Recovery." Report No. 18895-PH, Poverty Reduction and Economic Management Sector, East Asia and the Pacific Office. Washington, DC: World Bank.

Wright, Melissa. 2001. "Desire and the Prosthetics of Supervision: A Case of Maquiladora Flexibility." *Cultural Anthropology* 16 (3): 354–73.

Yates, Michael. 2003. *Naming the System: Inequalities and Work in the Global Economy.* New York: Monthly Review Press.

Yearbook of World Electronics Data Volume 3 2004/2005 Emerging Markets. 2004. Wantage, UK: Reed Electronics Research.

Yeung, Henry Wai-chung. 2002. "Industrial Geography: Industrial Restructuring and Labour Markets." *Progress in Human Geography* 26 (3): 367–79.

Yuen, E., and Hui Tak Kee. 1993. "Headquarters, Host-Culture, and Organisational Influences on HRM Policies and Practices." *Management International Review* 33 (4): 361–83.

Zuboff, S. 1988. *In the Age of the Smart Machine: The Future of Work and Power.* New York: Basic Books.

Index

Page numbers in italics refer to tables and figures

oligarchs, 54
O'Riain, S., 52
Osterman, P., 6

panoptic work regime, 1–2, 20–21, 22–23,
88; anti-union strategies of, 99, 149, 153,
157–58, 162; background investigations
and, 159–62; commitment and, 99, 174,
193–94, 199; disciplinary management
and, 125, 149; external labor control
and, 155–62; flexibility and, 125; home
visits and, 161–62; job fairs and, 148–
49, 159; prevalence in Philippines of,
213; recruitment and, 156–59; workers'
perspectives on, 187–94. *See also* Storage
Ltd.
Pascual, C., 178
Patricia (Discrete Manufacturing), 119, 120,
205, 207
Peck, J., 12, 59, 71, 137
Penley, L., 17–18
peripheral human resource (HR) work
regime, 20–21, 23, 100–12; commitment
and, 174–75; flexibility and, 125–26;
prevalence in the Philippines of, 213–14;
strategic localization and, 164, 175; recruitment and, 165–67; workers' perspectives on, 194–99. *See also* Integrated
Production
Philippine Economic Zone Authority
(PEZA), 25, 59–61, 134; anti-unionism
and, 142–44, 146, 153–54; commitment
to locators of, 152; institutional autonomy
and, 223; labor control and, 143, 146; labor disputes and, 144, 150; private zone
developers and, 151; privatization and
special economic zones (SEZs), 89, 101,
124, 148, 150–55, 176, 222–23; recruiting and, 154–55. *See also* export processing zones
Philippine Ecozone Association (PHILEA),
151, 154; local state collaboration with,
155; Philippine National Police collaboration with, 154
Philippine Electronics Industry, 43–51; casualization of labor in, 69; concentration of,
48, 51; employment in, 46, 47, 61; exports of, 44–46, 45, 61; growth rate, 46;
imports of, 49–50; investments in, 46,
60–61; labor force of, 65–69, 226; multinational corporations dominance of, 47–
49; nationality of firms in, 49; share of
electronics in total imports, 49; state investment policies and, 51–52; strikes and
lockouts in, 70; value-added in, 50
Philippines; employment in, 63–64; global

production and, 3, 19, 29; Labor Code of,
92, 160, 164, 176, 219; labor force in, 62;
labor force compared with other Asian
countries, 66; labor market in, 62–63, 63;
labor movements in, 64–65; Medium
Term Comprehensive Employment Plan
in, 71; poverty in, 63; Region IV of, 63;
unemployment rates by education level in,
63; wages in, 64, 71
Piore, M., 6
place, 4, 30
political apparatuses, of production, 9, 11;
of flexible accumulation, 12, 173, 216
Power Tech (pseud.), 79; labor relations at,
85–87, quality at, 82, productivity at, 83;
unionism at, 86, 146; wages, 79. *See also*
Allied-Power Group; despotic work
regime
Public Employment Service Office (PESO),
74, 139, 154–55, 169, 219
Purcell, J., 17, 221
purchased commitment, 21, 23, 111–12,
199. *See also* Integrated Production; commitment

quality, 7–9, 81–82. *See also* flexible production; high performance work organization

Ramon (Discrete Manufacturing), 116, 205,
207
Ramos, F., 44, 58, 145
recruitment, labor, 20–21, 133; Allied-
Power and, 137–39; Discrete Manufacturing and, 169–70; dispersed strategy of,
153, 159; export processing zones and,
137–38; gender and, 156–57, 176, 224;
ideal workers and, 156–58; Integrated
Production and, 165–67; local state involvement in, 141–42, 144, 154–55; Storage Ltd. and, 156–57. *See also* labor
market segmentation
regulatory state, 53, 58–59, 71, 75, 133,
218–20, 222–23, 226; institutional autonomy and, 223. *See also* state, role of the
Remulla, J., 86, 140–42, 145, 147
Rene (Discrete Manufacturing), 116, 205
reproduction, of social relations, 12
research methods, 24–28
restructuring, industrial, 4, 5, 11, 130
Roland (Discrete Manufacturing), 116, 205
Rosario Workers' Center (RWC—psued.),
183–84, 227; and women's issues, 184
Rose (Allied, Ltd.), 182
Rubery, J., 15, 163, 220